Sebastian Hellmann

Research and Technology Development on Superconducting Current Limiting Transformers

Karlsruher Schriftenreihe zur Supraleitung Band 023

HERAUSGEBER

Prof. Dr.-Ing. M. Noe
Prof. Dr. rer. nat. M. Siegel

Eine Übersicht aller bisher in dieser Schriftenreihe erschienenen Bände finden Sie am Ende des Buches.

Research and Technology Development on Superconducting Current Limiting Transformers

by
Sebastian Hellmann

KIT Scientific Publishing

Karlsruher Institut für Technologie
Institut für Technische Physik

Research and Technology Development on Superconducting Current Limiting Transformers

Zur Erlangung des akademischen Grades eines Doktor-Ingenieurs von der KIT-Fakultät für Elektrotechnik und Informationstechnik des Karlsruher Instituts für Technologie (KIT) genehmigte Dissertation

von M.Sc. Sebastian Hellmann

Tag der mündlichen Prüfung: 27. April 2018
Hauptreferent: Prof. Dr.-Ing. Mathias Noe
Korreferent: Prof. Dr. rer. nat. Steffen Elschner

Impressum

KIT Scientific Publishing

Karlsruher Institut für Technologie (KIT)
KIT Scientific Publishing
Straße am Forum 2
D-76131 Karlsruhe

KIT Scientific Publishing is a registered trademark of Karlsruhe Institute of Technology.
Reprint using the book cover is not allowed.

www.ksp.kit.edu

Print on Demand 2019 – Gedruckt auf FSC-zertifiziertem Papier

ISSN 1869-1765
ISBN 978-3-7315-0804-5
DOI 10.5445/KSP/1000083484

Acknowledgments

The outcome of the research work presented in this thesis would not have been possible to achieve without the support and supervision of Prof. Dr.-Ing. Mathias Noe (ITEP, KIT) and the whole Institute of Technical Physics.
I further want to thank Dr. Markus Abplanalp (ABB, Switzerland) and Dr. Lukas Hofstetter (ABB, Switzerland) for their continuous support and input for this research work. Additionally I want to thank Prof. Dr. Steffen Elschner and Mr. Andrej Kudymow for their interest in this project and for the countless hours of helpful technical discussion during its course towards completion.
For the professional support in the laboratory and in the workshop, during the manufacturing and the testing of the superconducting current limiting transformer, I want to thank Mr. Johann Willms, Mr. Uwe Walschburger, Mr. Uwe Mirasch and Mr. Holger Fillinger. And lastly, my biggest thanks go to all the colleagues and workmates I encountered during my time at the ITEP and who remain for me as friends, even after the end of my time at the KIT.

Karlsruhe, 8th March 2018

Sebastian Hellmann

To my grandparents

Kurzfassung

Moderne Energieübertragungsnetze unterliegen ständigen Veränderungen hin zu höherer Übertragungseffizienz, zu größeren Leistungsdichten in Ballungsgebieten und Innenstädten und hin zu einer zunehmenden Dezentralisierung der Energieeinspeisung auf den unteren Spannungsebenen. Die nötige Anpassung der Netztopologie an die genannten Punkte führt oft zu einer Reduktion der Netzimpedanz und zu einer Zunahme der Vermaschung des Stromnetzes.

Eine geringere Netzimpedanz führt in Fehlerfällen im Stromnetz jedoch auch zu einem Anstieg der Kurzschlussleistungen. Ebenso kann, bei einer dichteren Vermaschung des Stromnetzes, ein auftretender Fehler weitreichendere Konsequenzen haben und den sicheren und unterbrechungsfreien Netzbetrieb stärker gefährden.

Neuartige Netzkomponenten, basierend auf supraleitenden Materialien, können hier Abhilfe schaffen und durch signifikante Vorteile gegenüber konventionellen Netzkomponenten und Schutzeinrichtungen zu einem unterbrechungsfreien und verlässlichen Netzbetrieb beitragen sowie die steigenden Auswirkungen von Netzfehlern effizient eindämmen.

Die vorliegende Arbeit zeigt dies am Beispiel eines Transformators, der mit einer supraleitenden Sekundärwicklung ausgestattet ist. Die intrinsischen Eigenschaften des supraleitenden Materials erweitern hierbei den Transformator um die Fähigkeit, auftretende Kurzschlussströme im Stromnetz aktiv und effizient zu reduzieren. Im Vergleich zu konventionellen Netzschutzeinrichtungen geschieht dies ohne nennenswerte Verzögerung nach dem Auftreten des Fehlers.

In den Kapiteln dieser Arbeit wird das nötige Grundwissen zur Supraleitung, zur Funktionsweise und zur Konzeptionierung von supraleitenden strombegrenzenden Transformatoren dargestellt. Weiter wird das theoretisch erarbeitete Wissen im Bau eines Labordemonstrators der 1 MVA Leistungsklasse und dessen anschließender experimenteller Evaluierung angewandt.

Die einzelnen Abschnitte der vorliegenden Arbeit und deren Inhalt sind im Folgenden kurz zusammengefasst:

Kapitel 1 vermittelt die Grundlagen der Supraleitung und deren Anwendung in der Energietechnik. Hierbei wird speziell auf Zusammenhänge eingegangen, die für die Kon-

struktion und den Bau eines strombegrenzenden supraleitenden Transformators von Wichtigkeit sind. Das Kapitel wird abgeschlossen mit einer Literaturrecherche zum Stand der Technik von supraleitenden Transformatoren und zu weiteren Anwendungen der Supraleitung in der Energietechnik.
Im zweiten Kapitel der Arbeit wird der Auslegungsvorgang für supraleitende Transformatoren mit strombegrenzenden Eigenschaften erarbeitet. Hierzu wir ein Verfahren vorgestellt, mit welchem die zu erwartenden transienten Eigenschaften eines Transformatorentwurfs simuliert und optimiert werden können. Das zweite Kapitel endet mit der Vorstellung des erarbeiteten Designs für einen supraleitenden strombegrenzenden Transformator der 1 MVA Leistungsklasse, welcher im weiteren Verlauf dieser Arbeit als Labordemonstrator gebaut und experimentell untersucht wird.
Kapitel 3 beschreibt verschiedene experimentelle Untersuchungen an supraleitenden Bandleitern, die durchgeführt wurden um deren Eignung zum Einsatz in strombegrenzenden Transformatoren festzustellen. Die Untersuchungsergebnisse werden anschließend verwendet, um den Auslegungsvorgang für supraleitende Transformatoren zu optimieren und um einen geeigneten, kommerziell erhältlichen Supraleiter zum Einsatz in dem entworfenen supraleitenden Transformator festzulegen.
In Kapitel 4 wird der Aufbau des supraleitenden strombegrenzenden Transformators beschrieben. Hierbei wird auf die einzelnen Transformatorkomponenten im Detail eingegangen und es werden die konstruktiven Details des Aufbaus erläutert.
Im fünften Kapitel der Arbeit werden die experimentellen Untersuchungen an dem gebauten supraleitenden Transformator vorgestellt. Das Kapitel umfasst die Beschreibung der durchgeführten Messungen als auch der verwendeten Messmethoden und Laboraufbauten. Die beschriebenen Messungen behandeln die Effizienz der Strombegrenzung des Transformators im Fehlerfall, den Wirkungsgrad und die Verlustmechanismen des Transformatoraufbaus bei verschiedenen Belastungen, die AC-Verluste der verwendeten Supraleiter als auch die Effektivität des eingesetzten Kühlkonzepts.
Kapitel 6 stellt die Ergebnisse der durchgeführten Messungen vor und diskutiert diese im Detail. Hierbei wird gezeigt, dass der gebaute Transformator bei einem Kurzschluss den Fehlerstrom im Maximum der ersten Halbwelle auf 68 % des prospektiven Kurzschlussstroms reduziert und dieser im Verlauf der Strombegrenzung weiter, auf unter 30 % des prospektiven Wertes absinkt. Die durchgeführten Verlustmessungen zeigen eine Gesamteffizienz von 99.08 % für ein dreiphasiges Transformatormodell unter Volllast. Darin enthalten sind die gemessenen AC-Verluste des Supraleiters mit 633 W. Weiter wird die Bestimmung der relativen Kurzschlussspannung zu 2.89 % gezeigt.
Zusätzlich werden die Messergebnisse mit den, in Kapitel 2 erarbeiteten Simulations-

ergebnissen verglichen und zu einer weiteren Optimierung des vorgestellten Auslegungsvorgangs verwendet. Zwischen den Mess- und Simulationsergebnissen zeigt sich hierbei eine sehr gute Übereinstimmung, welche die Richtigkeit der verwendeten Simulationsmodelle bestätigt.
Abschließend wird das Gesamtergebnis der vorliegenden Arbeit zusammengefasst und es wird ein Ausblick auf zukünftig mögliche Forschungsarbeiten zum Einsatz von supraleitenden strombegrenzenden Transformatoren aufgezeigt.

Abstract

Modern power grids undergo continuous changes towards higher efficiency in power transmission, increased available power density in urban centers and highly populated areas and towards higher decentralization of the energy production on the low-voltage levels. The necessary adjustment of the grid topology often results in a reduction of the grid impedance and further in an increasing number of grid interconnections.

Disregarding the desired positive effect of this change, a lower grid impedance also leads to an increase of the short-circuit power in case of a fault in the power grid. Appearing faults can have more severe and extensive consequences, enabled by the increased number of interconnections in the grid. This can pose a threat to a reliable and uninterrupted grid operation.

Novel grid appliances, based on superconducting materials, can help to reduce these threats and offer significant advantages in comparison with conventional grid components. With this, they can contribute to a reliable and fail-safe grid operation even with a reduced grid impedance.

This thesis introduces a hybrid transformer design with a normalconducting primary and a superconducting secondary winding. In this design, the intrinsic properties of the superconducting material extend the conventional transformer by the ability to actively and efficiently limit short-circuit currents in a power grid. In comparison with conventional grid protection methods, the superconducting current limiting transformer limits fault currents instantaneously.

This thesis gives an introduction to modern superconductor materials, to the functionality of a superconducting current limiting transformer and to the design process of such a device. The acquired knowledge is then applied to manufacture and test a superconducting transformer demonstrator of the 1 MVA-power-class.

The different sections of this work are briefly described in the following:

Chapter 1 introduces the basics of superconductivity and the limiting factors for the utilization of high-temperature superconductors in power devices. The focus lies on the application in a superconducting current limiting transformer. The chapter concludes with a state-of-the-art overview of superconducting transformers and other superconducting power applications.

The second chapter introduces the design process behind a superconducting current limiting transformer as well as a method for the prediction of the transient operational characteristics of a superconducting transformer design. This chapter ends with the generation of a design for a transformer of the 1 MVA-class, which is manufactured and tested during the further course of this thesis.
Chapter 3 describes preliminary experiments on superconducting tapes, carried out to determine their suitability for the application in a current limiting transformer design. The experimentally acquired results are used to optimize the transformer design process and to determine a commercially available superconducting tape for the utilization in the test setup of the developed superconducting transformer demonstrator.
In chapter 4, the manufacturing of the designed superconducting transformer demonstrator is described in detail and constructional feature of the different transformer components are explained.
Chapter 5 presents the experiments conducted on the manufactured superconducting transformer demonstrator. This includes the description of the experimental setups in the laboratory, the applied measurement methods as well as a listing of the utilized experimental equipment. The measurements focus on the performance of the transformer demonstrator during a current limitation, the efficiency of the full transformer assembly, a loss analysis on the different transformer components as well as on the AC-loss contributions of the superconducting material.
The results of the experiments are presented and discussed in detail in chapter 6. It is shown that the superconducting transformer reduces a fault current in the peak of the first half wave to 68 % of the prospective current. The current further decreases to less than 30 % in the peak of the sixth half-wave. The measurements of the transformer losses show an efficiency of 99.08 % for a three-phase transformer model. In this value, the AC-losses in the superconductors are considered with 633 W under nominal load. For the short-circuit impedance a value of 2.98 % is determined.
This chapter also presents a comparison between the measurement results and the results of the transient simulations acquired during the transformer design process. It is shown that the simulated values agree well with the measurement values. This verifies the applied simulation models, used during the transformer design process.
The thesis concludes with a summary of the acquired results and gives an outlook on possible follow-up research projects, necessary to further investigate on the significant potential of superconducting current limiting transformers in future power grids.

Contents

1 Introduction

1.1 Motivation and Scope of Work

Over the last decades the electric power grids of many countries all over the globe experienced significant structural changes. Among others, these changes in the grid structure are driven by the following economic, ecological and political factors:

- more efficient power transmission over longer distances
- higher power densities in city centers and other highly populated industrial or residential areas
- increasing amount of renewable sources of energy

A higher efficiency of long-length power transmission can be achieved by reducing the impedance of the transmission line. This can be done by increasing the voltage level the line is operated at. The voltage levels in transmission lines reach nowadays up to 1 MV and even beyond. A higher efficiency for the transmission of electric power can as well be achieved by reducing the electrical resistance of the transmission line. This also applies for transformers on both termination points of a high-voltage transmission line.
The reduced over-all inductance of the transmission line has the effect of lower losses per unit of transferred energy. However, a lower impedance in a long-distance electric transmission line also means that occurring short-circuit currents in the power grid can significantly increase in amplitude and consequent destructive potential.
The increasing demand for higher power densities in urban centers and in highly populated areas can also be fulfilled in two different ways, by either lowering the over-all impedance of the power grid in the low-voltage distribution level or by increasing the number of interconnections in the grid structure. This increases the power density but also leads to an increase of the short-circuit power.
The third factor for the changing grid structure is the incorporation of an increasing number of renewable sources of energy. This proves to be especially challenging for certain types of renewable sources, which can experience a high fluctuation in the gross energy production. Such renewable sources are wind and photovoltaic energy (PV).

For example, in February 2018 the total installed electric power generation in Germany, including all types of power plants, added up to 207.9 GW [Fra18]. More than 43.4 % (90.3 GW) of the total installed electric power were renewable sources in the form of wind and solar power, in which especially the solar power installations and smaller wind farms are connected to the grid on the low-voltage level. The high volatility of wind and PV-energy generation can lead to challenging power flow scenarios in the power grid, caused by a mismatch between energy production and demand.

This problem is increased by the fact that the power grids in most countries were originally designed and built in the first half of the last century. Back then, a power grid included a small number of larger power plants. These power plants were connected to the high or medium-voltage distribution grid. The average private or commercial consumer was located in the low-voltage level distribution grids. The grids were designed to experience a power flow in one direction - from the power plant to the consumer of electrical energy.

To deal with the possible reversal of the power flow, it can be useful to increase the number of interconnections between different branches of the low-voltage distribution grid. This leads to an efficient and more direct load distribution during times of higher energy production from PV-installations or wind farms. However, the secondary effect of an increased number of interconnection in the low-voltage distribution grid is a significant increase in short-circuit power and a subsequently increased danger of damages to grid installations.

These three changes to the grid structure have in common that the increase in short-circuit power is caused by a reduction of the grid impedance.

The higher short-circuit currents can be countered with conventional grid protection methods, as, for example more powerful circuit breakers and load disconnectors, in-series shunt reactors or by the use of transformers with artificially increased short-circuit impedance. However, all conventional methods for the reduction of short-circuit currents come either with additional investments into higher rated switchgear or protection equipment or, they are based on an increase of the grid impedance or by inserting additional inductive elements into the power grid. These methods are either increasing the investment into building and maintaining the power grid or, they decrease the overall efficiency of the power transmission.

During the course of this thesis the functional principle of a superconducting current limiting transformer is introduced as a novel concept for addressing increasing short-circuit currents in power grids. The project described in this work, features a superconducting current limiting transformer based on commercially available superconductors to replace the conventional, copper or aluminium based conductors in the secondary winding of a 1 MVA-class distribution transformer.

Scope of Work

Since the effective short-circuit current limitation with conventional means is often based on increasing the grid impedance in normal operation, the fault current limitation with superconductors provides a valid alternative. The scope of the work presented during the following chapters of this theses, is aiming at the general technology development for the design, construction and manufacturing process of a superconducting fault current limiting transformer. The goal of the project is to develop and validate a design process by manufacturing and testing of a prototype for a superconducting current limiting transformer.

The full scope of work of this thesis is divided into the following main topics:

- Develop a design procedure for current limiting transformers, enabling the mechanical and electrical design process of the superconducting and normalconducting transformer parts as well as of the passive transformer components. This procedure allows the prediction of normal and fault current operation behavior. The design procedure includes the optimization of the main transformer parameters for a specific transformer design.

- Evaluate commercially available conductor materials for their application in current limiting transformers. This covers general properties of the superconducting tapes as well as specific properties, such as short-circuit current limitation capabilities, the quench and quench propagation behavior as well as an estimation for the recovery-under-load capabilities.

- Design and manufacturing of a transformer prototype. The prototype should feature at least one superconducting winding, which provides resistive current limitation properties. The superconducting winding is based on a simple conductor concept. The manufactured prototype of a 1 MVA power-class, represents a commonly used distribution transformer size.

- Tests of the manufactured prototype and comparison with simulation results. Of main interest are the total losses of the transformer, current limitation characteristics, the performance of the cryogenic system, the overall transformer efficiency and the AC-losses in the superconducting material. The test results are used to evaluate and improve a simulation tool, developed during the transformer design phase.

1.2 High-Temperature Superconductor Materials

Superconductors can offer a significantly reduced electrical resistance for all types of AC-power applications. Further, superconductors can provide much higher current densities in comparison to conventional conductor materials. This makes the use of superconducting materials especially interesting for increasing the efficiency by reducing Joule-heating losses, or for constructing more compact and lightweight power devices, such as motors, transformers or generators.

Among the many materials and material compositions that are classified as superconductors [BK04, WMH15, MGC63], only a few are used for technical applications. Until the mid-nineteen-eighties, all discovered superconductors were, so-called, *low-temperature superconducting materials* (LTS-materials) which require temperatures below 23 K to reach the desired superconducting state [BK04, Kom95]. The energy-intense cooling with liquid helium makes the use of low-temperature superconductors for AC-applications economically non-viable for most of the cases.

With the discovery of superconducting properties for metal-oxides, by J. G. Bednorz and K. A. Müller in 1986, a new class of superconducting materials was found [BM86]. One year later, superconducting properties were confirmed for $YBa_2Cu_3O_{7-x}$, a yttrium-barium-copper-oxide material composition, which requires an operating temperature of below 94 K to be in the superconducting state [WAT$^+$87, BK04]. This temperature can be provided by simple and economic cooling, with liquid nitrogen (LN_2) as a coolant. Liquid nitrogen at an ambient pressure of 1013 mbar has a boiling point of 77.3 K and is widely used for cooling $YBa_2Cu_3O_{7-x}$-superconductors in open bath cryostats.

Nowadays $YBa_2Cu_3O_{7-x}$ is by far the most important high-temperature superconducting material (HTS-material) for AC-power applications. It is often abbreviated with YBCO and especially suited for the application in superconducting fault current limiters [Kal11, Mel12, Sei15, NS07, HEB$^+$12, EKB$^+$12].

The rare-earth element yttrium (Y) in the $YBa_2Cu_3O_{7-x}$-material composition can also be replaced by other rare-earth metals. The resulting superconductor is therefore also referred to as *Re*BCO, with the *Re* standing for the rare-earth metal component.

The fundamentals of superconductivity together with a detailed description of the different types of superconducting materials in the low-temperature as well as in the high-temperature range, is covered well in the literature [BK04, See98, Kom95]. Therefore, the explanations on the basics of superconductivity, given in the following sections of this chapter, are focused on the properties of *Re*BCO-superconducting material, as it will be used during the course of this project. The basic material properties of the *Re*BCO-material are essential to understand the functional principle of a superconducting current limiting transformer and the associated design procedure.

Critical Current Density in Superconducting Materials

The maximum transport current in a superconductor is significantly higher compared to conventional conductor materials of the same cross section. However, the transport current in a superconducting material is limited. This maximum transport current is referred to as the *critical current*, I_c and it is one of the most important characteristics of superconducting materials used in power applications.

For comparison, the critical current is usually stated as a critical current density J_c, which is given as the critical transport current of the superconductor I_c, divided by the superconductor cross section area A_{sc}.

The typical electric field versus transport current behavior ($E(J)$-behavior) of a high-temperature superconductor is depicted in Figure 1.1. The figure shows that the electric field E along the superconductor is zero for small current density values. If the current in the superconductor is increased above the critical current density J_c, the electric field rises sharply and follows a power-law. For many applications of high-temperature superconductors, a threshold electric field E_c, of 1 µV per cm superconductor length is used to define the critical current.

For *Re*BCO-materials the voltage drop depending on the current density in the superconductor follows the power-law stated in Equation 1.1.

$$E = E_c \left(\frac{J}{J_c} \right)^n \tag{1.1}$$

Here, E represents the voltage drop along the superconductor sample as an electric field, whereas E_c represents the criteria of 1 µV / cm. The current density J and J_c stand for the actual current density and the critical current density respectively, for the specified criteria E_c. The exponent for current *Re*BCO-materials is usually in the range of 20 to 40 [Haz17]. The large exponent in Equation 1.1 leads to the sharp rise of the voltage drop, seen in Figure 1.1, for values higher than the critical current density.

For current values above the critical current, the resistance of the superconductor increases. The Joule-heating in the superconductor and the superconductor temperature increase consequently, and a thermal runaway of the superconductor material is possible. This transfer of a superconductor from the superconducting to the normalconducting state is also referred to as the *quench* of the superconducting material.

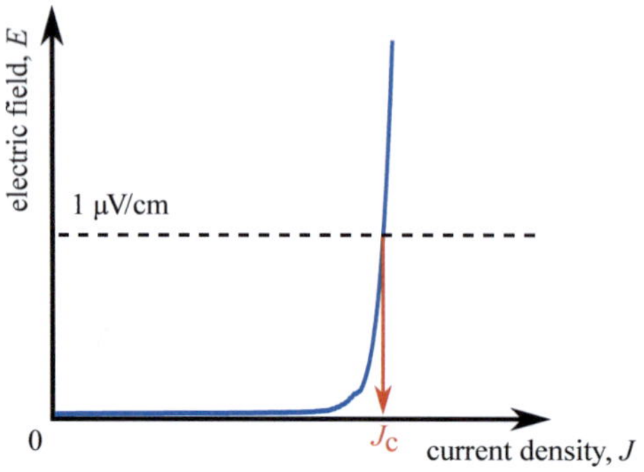

Figure 1.1: The critical current density of a superconductor is specified via the voltage drop along the superconductor. Often, a threshold value of 1 µV voltage drop per centimeter superconductor length is used to define the end of the superconducting regime and therefore the critical current J_c (not to scale)

Critical Temperature of Superconducting Materials

Figure 1.2 shows the resistance versus temperature behavior ($R(T)$-behavior) of common high-temperature superconductor materials. It can be seen, that above a so-called *critical temperature* T_c, the superconductor material is normalconducting and shows a linear dependency on the temperature (blue line), as it can be observed for many metal-based normalconductors. For temperatures beneath the critical temperature, the resistance of the superconducting material sharply decreases and can be assumed to be zero for DC-currents. The critical temperature, T_c is considered a second limiting property for the operation of high-temperature superconducting materials and is specific for each superconducting material composition.

The critical temperature for *Re*BCO-materials can be determined by fully immersing a superconductor sample in an open bath of liquid nitrogen at the boiling point of 77.3 K. Thereafter, the sample temperature is slowly increased, while the sample resistance is continuously measured and recorded together with the measured temperature of the sample. Guaranteeing a homogeneous temperature distribution inside the sample material is crucial for the precise determination of the critical temperature using this measurement procedure. The literature describes techniques to avoid strong temperature gradients inside the superconductor sample during temperature dependent measurements [Eki06, Ber11].

To measure the resistance of the superconductor sample, a precise sample current, significantly smaller than the critical current density of the material, is continuously applied to the sample.

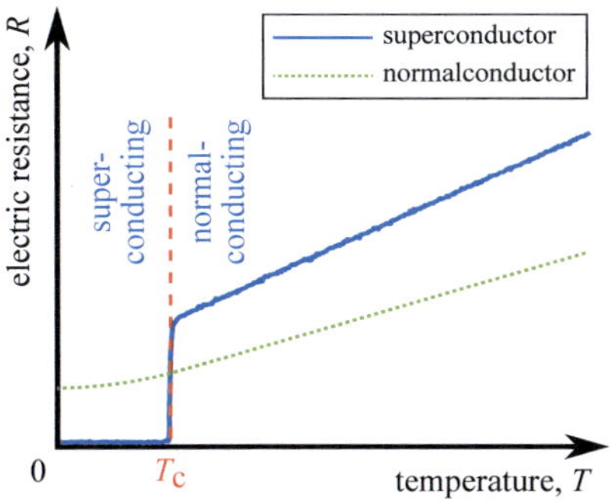

Figure 1.2: The sketch shows the resistance of a superconducting material versus the temperature T. The critical temperature T_c indicates the transit between the normalconducting regime and the superconducting regime of the material. The green line represents the temperature dependent resistance of a normal conducting metal (not to scale)

Influence of External Magnetic Fields

The third factor, considerably affecting the state of a superconducting material, is the applied external magnetic field density, $B_{ext.}$. An external magnetic field reduces the critical current density J_c of a superconductor as well as the value of the critical temperature T_c. For external magnetic fields higher than the *critical external magnetic field*, $B_{ext.,c}$, the superconductor fully leaves the superconducting state and becomes normalconducting.

The measurement results in Figure 1.3 visualize the magnetic field dependence of the critical current density of a typical *Re*BCO-thin-film material at a temperature of 77.3 K. It can be seen, that not only the external magnetic flux density, but also the angle of the external magnetic field, α has a significant influence on the critical current density of the superconductor material. The external magnetic field vector $\vec{B}_{ext.}$ is perpendicular to the vector of the transport current in the superconductor sample, $\vec{I}_{transp.}$. The field angle α in the graph is specified as the angle between the external magnetic field vector and the line perpendicular to the surface of the superconductor sample, as it can be seen from the sketch in Figure 1.3.

The strong variation of the relative reduction in the critical current density versus the angle of the external magnetic field is characteristic for *Re*BCO-thin-film materials. The dependency on the magnetic field angle can change for different manufacturing processes for *Re*BCO-thin-film superconductors.

The measurement results in Figure 1.3 show that even for relatively small magnetic fields of 50 mT, as they can be present in superconducting transformer windings, the relative critical current density decreases to less than 60 % of the original value. Therefore, the dependence of the critical current density on the external magnetic field is an important

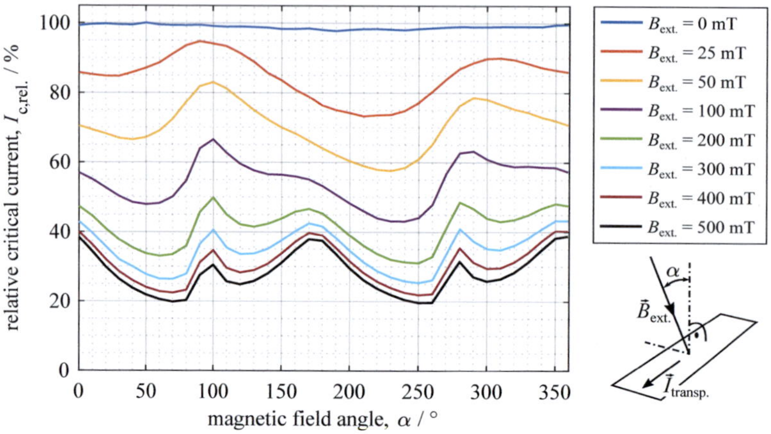

Figure 1.3: Critical current density of a *Re*BCO-thin-film material, depending on the external magnetic flux density $B_{ext.}$, the angle between the magnetic field and the perpendicular line of the material surface. The critical current density is normalized to the self field current density of the superconducting material. The measurement was conducted at a temperature of 77.3 K (thanks to Mr Aurélien Godfrin for providing the measurement data presented in this figure)

parameter in the design process for superconducting coil and magnet arrangements.

Combination of Limiting Factors in Superconducting Materials

For many applications of *Re*BCO-superconductor materials, the critical current density, the critical temperature and the critical external magnetic field, as described in the previous sections, are important limiting factors.

Figure 1.4 shows the operational limit of a superconducting material, defined by the value of the current density J, the external magnetic flux density $B_{ext.}$ and the temperature T. The operational limit in the presented sketch is visualized as a non-planar surface in green color. In the graph, all combinations of values for J, T and $B_{ext.}$, that result in an operational point inside the volume underneath this surface are in the superconducting regime of the material. One exemplary combination of values is indicated by red dashed lines.

Due to the complex correlation between the current density, the temperature and the external magnetic field, it is important to precisely evaluate a superconducting material before designing possible applications. Especially for operating conditions where the limiting factors can vary during operation, as for example in superconducting transformers, motors, generators or cables which are operated under different load conditions or with changing environmental factors. The closer the material in the foreseen application

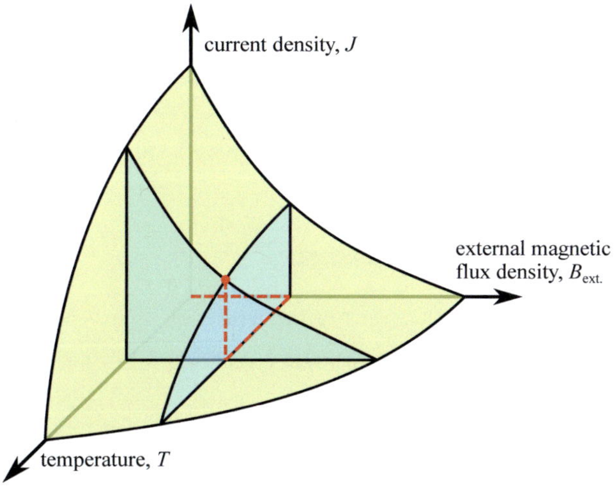

Figure 1.4: Visualization of the limiting factors for the superconducting state of a superconductor, defined by the current density, J the external magnetic flux density, $B_{\text{ext.}}$ and the temperature, T. Every combination between J, $B_{\text{ext.}}$ and T, beneath the surface visualized in green color, is inside the superconducting state (not to scale)

is operated at the operational limit, shown in Figure 1.4, the more important it is to rely on precise measurement values for the critical current density, the critical temperature and the influence of the external magnetic field on the superconductor material.

AC-Losses in *Re*BCO-Materials

When transporting AC-current or when exposed to a changing magnetic field, two different loss mechanisms occur in a HTS-conductor - the so-called *hysteresis losses* and the *eddy-current losses*. The loss mechanism behind the hysteresis losses is based on the dissipation of energy due to the changing magnetic flux density of magnetic fields, penetrating the superconducting material. The eddy-current losses develop in the normal conducting layers (mainly stabilization and substrate) of the superconductor tape composition. The eddy-current losses in a superconducting tape are identical to the eddy-current losses observed in conventional power applications.
Since both loss mechanisms only appear for AC-currents or a changing magnetic field inside the superconducting material, they are referred to as *AC-losses*.
The AC-losses of superconducting tapes in AC-power applications are usually small in comparison to the Joule-heating losses of conventional conductor materials. But still, for most applications even small AC-losses have to be divided by the efficiency of the cryogenic cooling equipment, and can therefore evolve into a significant factor for the total efficiency of superconducting AC-power applications.

Apart from a few analytic approaches for simple conductor geometries [Nor70, BI93, MW02], the calculation of AC-losses is often complex and can only be done by means of numerical simulations [Gri04, AMBT97, AMBM98, JAN+06]. This especially applies for complex geometries such as superconducting coils for transformer windings [PSJG15, GZP+14].

Current-Triggered Resistance Increase

If one or more of the critical values of a superconductor is exceeded during operation, the resistance of the superconductor increases following a power-law, as it is shown in Figure 1.1.

This intrinsic effect of the superconducting material is utilized in resistive fault current limiters, in which a superconducting element is connected in series to a load in a power grid, as shown in Figure 1.5. The circuit diagram shows the components of the grid, the short-circuit being indicated by a red arrow. This short-circuit connection fully or partially bypasses the nominal load and therefore causes a short-circuit current I_{fault}, significantly larger than the nominal operation current of the power grid.

The superconducting element is operated below its critical temperature. As long as the nominal current in the power grid is lower than the critical current of the superconducting material, the element is in the superconducting state and has no influence on the current flow in the power grid. The dimensioning of the critical current for the superconducting element is hereby done in such a way that any possible normal operation current is lower than the critical current. However, a fault current caused by a short-circuit causes a current in the superconducting element that exceeds the critical current of the superconducting material. The resistance of the superconducting element consequently rises, what in turn lowers the fault current in the grid by a resistance increase and therefore protects the power grid and grid installations.

The functional sequence of a fault current limiter based on superconducting *Re*BCO-material is shown in Figure 1.6. The upper graph in the figure shows the current through the superconducting element and the temperature of the superconducting material in the bottom graph. The green background color in both graphs indicates the duration of the fault in the power grid.

The top graph of Figure 1.6 shows that the current is not cut off after the fault but continues to flow through the superconducting element. If the efficiency of the cooling for the superconductor is higher than the Joule-heating inside the superconducting material, the superconductor is capable of recovering to the superconducting state even while carrying the nominal current. This capability is referred to as *recovery-under-load* (RuL)

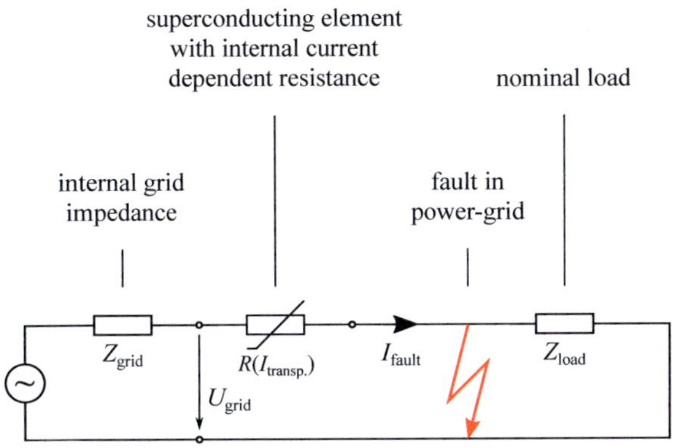

Figure 1.5: The circuit diagram shows a superconducting element installed for short-circuit current protection in a power grid. The superconducting material is used as a current-triggered resistive fault current limiter

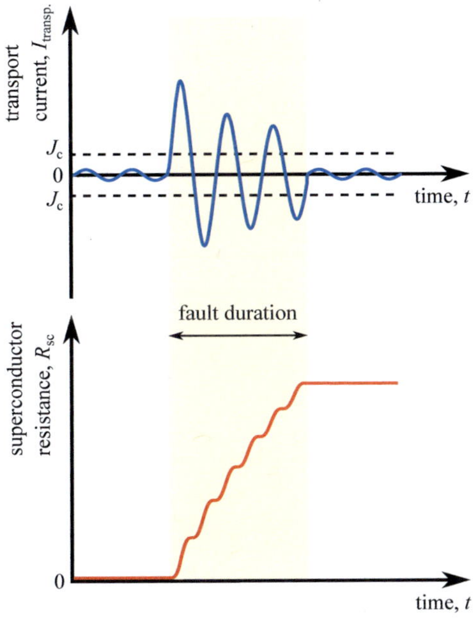

Figure 1.6: The upper graph shows the exemplary limitation of a fault current in a electrical grid due to the current-triggered resistance increase of the superconducting *Re*BCO-material. The lower graph depicts the increasing resistance, also referred to as the *quench* of the superconductor

of a superconducting fault current limiter.

Current-triggered resistive fault current limiters were among the first applications for high-temperature superconducting materials and their development is well documented in the literature [BBW+05, Sch09, EKB+12]. In addition to the pure resistive superconducting fault current limiters, there is also a variety of other types and methods for reducing appearing fault currents in a power grid, based on the application of superconducting materials [NS07, SBE+16, EBW+07, XGS+13, PLR+97].

1.3 *Re*BCO-Materials for Power Applications

*Re*BCO-materials, as introduced in Chapter 1.2, are brittle compounds with mechanical material properties similar to those of ceramics. To use these materials in power applications, it is necessary to manufacture the brittle *Re*BCO-material into conductor compositions that provide sufficient mechanical, thermal and electrical stabilization.
The most common design of commercial superconductors on the basis of *Re*BCO-materi-as is presented in Figure 1.7. The visualization shows a model of a thin-film *Re*BCO-supercon- ducting tape in a typical layered structure.
A superconductor tape is produced by first depositing thin buffer layers onto a metallic substrate material such as Hastelloy® or stainless steel. The substrate material primarily serves as a mechanical support for the whole tape structure and has a typical thickness of 20 µm - 100 µm. The deposited buffer layers enable the crystal growth of the *Re*BCO-material layer and chemically isolate the superconductor from the substrate material. The typical thickens of the buffer layers is less than 1 µm.
Secondly a thin, quasi-mono-crystalline layer of *Re*BCO-material is deposited on top the buffer layers. The *Re*BCO-layer has a usual thickness between 1 µm and 3 µm.
To improve the mechanical stability and to compensate for small, local defects in the critical current density of the *Re*BCO-layer, a protective metallic cap layer is added on top of the *Re*BCO-material during a third production step. For chemical reasons, this protective layer usually consists of silver and has a typical thickness between 1 µm and 3 µm.
Most applications require further electrical, thermal or mechanical stabilization of the tape material, to protect it in case of the loss of superconductivity or against mechanical damages. This can be achieved by electroplating or ion-sputtering additional layers of conductive material around the described tape structure, as shown in Figure 1.7. Depending on the intended application of the superconducting tape, the additional stabilization layer can have a thickness of more than 100 µm and often consists of copper.
A detailed description of the different production techniques for superconducting thin-film *Re*BCO-tapes can be found in the literature [Mel12, Sei15].
Figure 1.8 shows the microscopic view of the cross section areas for two commercially available superconducting *Re*BCO-tapes. The total average thickness of the electroplated

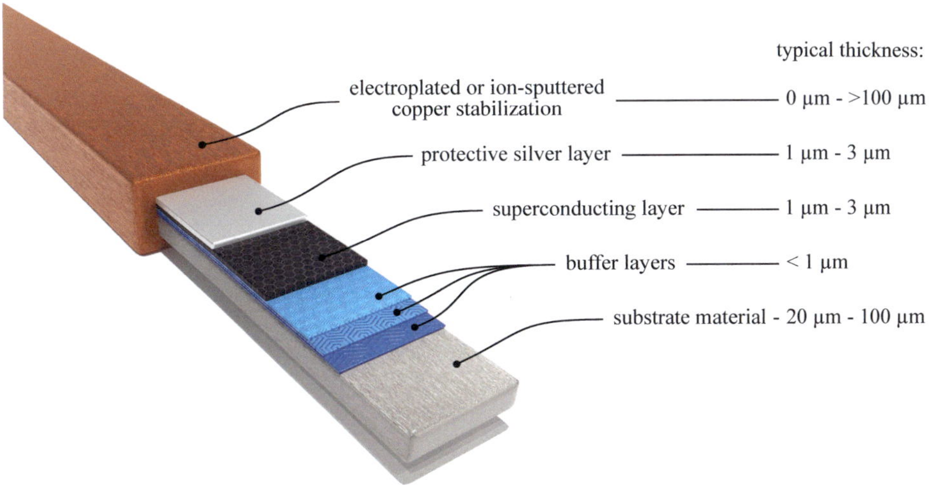

Figure 1.7: The figure shows the layered composition of a superconducting tape based on thin film *Re*BCO-material (artistic visualization, not to scale)

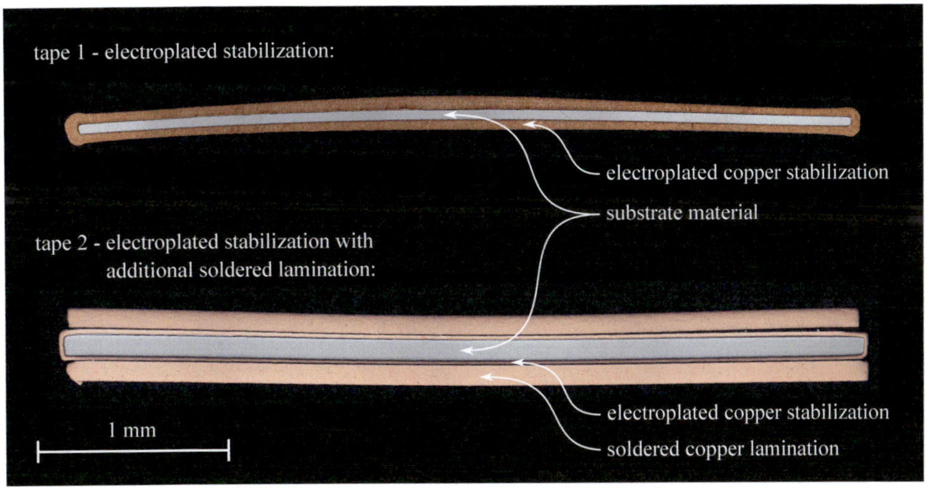

Figure 1.8: Optical microscopy image of the cross section area of two commercial superconducting tapes. The protective silver layer, the superconducting layer and the buffer layers are too thin to be visible in this image (thanks to Ms Brigitte Runtsch for providing the microscopy images presented in this figure)

copper stabilization on tape 1 is 105.3 µm. Tape 2 is stabilized by a electroplated copper layer with a total average thickness 40 µm and by two additional layers of soldered-on copper laminate with a total thickness of 185.5 µm. Due to the large scale of the cross section views it is not possible to distinguish the silver stabilizer, the superconductor layer or the buffer layers in the microscopy image of the superconducting tape compositions. Due to the low thickness of the material layers in the structure of superconducting *Re*BCO-tapes, the resulting conductor composition shows a good flexibility and can be bend in radii of down to less than 10 mm [Bar13, ZYS^{+}15], with the bending axis being parallel to the surface of the tape structure. This allows a relatively easy use of thin-film *Re*BCO-superconducting tapes for composing simple superconducting transformer windings, coils and magnets or also for the application in superconducting cables.

Commercial Availability of *Re*BCO Thin Film Tapes

For the use of superconducting *Re*BCO-materials in power applications, it is important that conductor materials are commercially available in long length, with sufficient material performance, with an economic cost to critical current density ratio and with a consistent production quality.
Superconductor manufacturers can nowadays provide *Re*BCO-superconducting tapes with a single piece length of up to 1000 m [Moo14]. This is especially of importance for the manufacturing of superconducting coils, magnets or in superconducting cables, for which it is often not possible to splice several shorter pieces of superconductor material into one large piece.
With present production techniques, it is further possible to manufacture long-length superconducting tapes with critical current densities of more than 500 A per mm^2 of the engineering cross section area [XZG^{+}17]. The engineering cross section area refers here to the cross section area of the whole tape composition and not only to the area of the superconducting *Re*BCO-material by itself.
Nowadays, the conductor materials are available at a price-performance ratio of less than 150 US$ per meter and kA of critical transport current [Moo14].
In the last 15 years, the production of superconducting *Re*BCO-tapes has made permanent progress towards longer available piece length and higher engineering current densities. Today, more than ten commercial suppliers are in business worldwide, offering different types of superconducting *Re*BCO-tapes.
The commercial tapes can be customized for the critical current density, electrical and thermal stabilization and other parameters, to meet the demands of the application. Many suppliers also offer a possible lamination of the superconducting tapes with additional layers of copper, brass or stainless steel tapes. This allows for the adjustment of the tape properties and is especially useful for power applications, current limiters and superconducting cables.
However, the production capacity of the commercial suppliers worldwide is still on a

pilot production level and the output of material is not sufficient.

1.4 State of the Art — Superconducting Current Limiting Transformers

Studies on superconducting power transformers started in the 1960s with the increasing availability of long-length low-temperature superconducting wires on the basis of NbTi and Nb3Sn compounds [JMMW96]. The outcome of these studies concluded that, without an significant increase of the cooling efficiency for the liquid helium refrigeration, a reduction of the AC-losses in the superconducting materials or an increase in the superconductors critical temperatures, this technology was not suitable for the efficient use in power transformers [JMMW96].
Until the beginning of the 1980s, the efficiency of the available refrigeration systems for the application of LTS-materials had improved and the manufacturers of LTS-wires achieved reductions in AC-losses. This made the application of low-temperature superconductors again interesting for commercial use in power transformers and resulted in the first superconducting transformer build by a cooperation between the Laboratoires de Marcoussis and Alstom in France in the year 1986 [FTLB88].
A collection of developed and tested power transformers on the basis of LTS-materials is listed in Table 1.1.
But even with reduced AC-losses in the LTS-wires and with higher efficiency of the refrigeration systems for the cooling, the applicability of the LTS-technology for power transformers remained questionable. One of the reasons for this is also, that the LTS-wires in the constructed prototypes were often not able to handle fault currents and therefore required external fault protection for the superconducting transformers. The difficulty with the quench stability arises from the fact that the specific heat of the LTS-materials significantly decreases for lower temperatures. This in turn, significantly reduces the quench stability, in comparison to the HTS-tapes presented in Chapter 1.3, and therefore requires external means of fault protection to circumvent thermal damages of the transformer windings.
The discovery of HTS-materials in 1986 instantly encouraged several research projects on the application of high-temperature superconductors for building superconducting power transformers. This increased interest was based on early research work, which suggested that the use of HTS-material over the use of LTS-wires could increase the efficiency of a power transformer by 35 % as well as lead to significant reductions in transformer size and weight [JMMW96].

Table 1.1: Listing of transformer projects based on LTS-materials (this list does not claim to be complete)

company / organization	year of publ.	country	rated power (kVA)	prim. voltage / sec. voltage (V)	prim. current / sec. current (A)	SC mate-rial	reference
Kyushu Univ.	1987	J	72	1057 / 218	68,4 / 332,9	NiTi	[MKMK88]
GEC-Alstrom	1988	F	80	660 / 1040	124 / 77	NiTi	[FTLB88]
Toshiba	1991	J	30	100 / 100	300 / 300	NiTi	[YTMI91]
Nagoya Univ.	1991	J	100	6600 / 210	15,2 / 476	Cu / NiTi	[KOH$^+$91]
Kyushu Univ.	1992	J	1000	3300 / 220	303 / 4545	NiTi	[FIT$^+$92]
ABB	1993	CH	330	6000 / 400	56,7 / 830	NiTi	[HABK93]
Osaka Univ.	1995	J	40	460 / 115	50 / 200	NiTi	[IMM$^+$95]

Table 1.2 lists a compilation of significant transformer projects based on HTS-materials. The first presented transformer prototype was manufactured by the Kyushu University in cooperation with the Fuji Electronic Co. and Sumitomo Electric Industries in the year 1997 [FIT$^+$97]. The transformer featured two fully superconducting windings, build from Bi2223-material, also referred to as *BSCCO-material.* It was tested at operating temperatures from 77.3 K down to 66 K with a corresponding rated power of 500 kVA and 800 kVA respectively. Including the refrigeration system, the transformer showed a total efficiency of 99.1 % at 77.3 K and 99.3 %, when operated with sub-cooled liquid nitrogen at 66 K.

In 1998, ABB developed a superconducting three-phase transformer with a Bi2223-based winding concept [Zue98]. This transformer was operated at 77.3 K and successfully tested in the power grid over the course of one year. The published results suggested an economical break-even point for transformer designs larger than 40 MVA.

In the year 2003, Siemens presented a superconducting transformer for railway applications [SSLM03]. This was one of the last large-scale transformer projects based on BSCCO-materials, that was reported in the literature.

In 2007, the University of Nagoya (Japan) presented measurement results on a single phase 33.3 kVA superconducting transformer prototype fully based on *Re*BCO superconducting tapes. The transformer windings were fully manufactured from *Re*BCO-thin-film

Table 1.2: List of manufactured HTS transformer prototypes (this list does not claim to be complete). The minus-sign for the fault current limitation indicates no current limitation, a plus-sign indicates current-limitation and two plus signs indicate current limitation with additional recovery-under-load capabilities

company / organization	year of publication	country	rated power (kVA)	pri. / sec. voltage (kV)	pri. / sec. current (A)	SC-material	operat. temp. (K)	fault current limitation	literature reference
Kyushu Univ. (Fuji, Sumitomo)	1997	J	800 / 500	6.6 / 3.3	76 / 152	Bi2223	66 / 77.3	-	[FIT+97]
ABB (EDF, ASC)	1998	CH	630 (3-ph.)	18.72 / 0.420	19.9 / 886	Bi2223	77.3	-	[Zue98]
Waukesha (ORNL, IGC,RG&E)	2002	USA	1000	13.8 / 6.9	72.5 / 145	Bi2212	30	-	[SMWJ02]
Siemens (VSH)	2003	DE	1000	25 / 1.5	40 / 720	Bi2223	66	-	[SSLM03]
University of Nagoya	2007	J	33.3	6.6 / 0.21	8.7 / 275	*Re*BCO	77.3	+	[OKI+07, OKH+09]
Karlsruhe Institute of Technology (KIT)	2011	DE	60	1 / 0.6	60 / 100	*Re*BCO	77.3	++	[BNK11, Ber11]
University of Nagoya	2011	J	666.7	22 / 6.6	52.5 / 175	*Re*BCO	77.3	++	[HKK+11, HKHO11]
Callaghan Innovation	2013	NZ	1000 (3-ph.)	11 / 0.415	30 / 1390	*Re*BCO	70	-	[GSJA13, GSB+11, SGP+12]

tapes. This transformer is the first introduced design, that enabled active fault current limitation by making used of the current-triggered resistance increase of the superconducting *Re*BCO-material. To adjust the quench and current limitation characteristics to the desired performance, the secondary winding of the transformer was composed of two types of *Re*BCO-tapes with different thermal and electrical stabilization [OKI+07, OKH+09]. The first design for a prototype of a superconducting fault current limiting transformer with recovery-under-load capabilities was presented by the Karlsruhe Institute of Technology in 2010 [BNK11, Ber11]. This transformer had a complex nominal power of 60 kVA and successfully limited a short-circuit current to 52 % of the prospective current in the first half-wave. After the fault was cleared from the electric grid, the *Re*BCO-tape in the transformer winding assembly proved to be able to recover from the quenched state (186 K) back to the superconducting state in 2.3 seconds, while carrying the full nominal transformer load. The transformer prototype had a conventional iron core and a copper primary winding, both operated at 77.3 K.
The University of Nagoya presented in 2011 a further improved laboratory prototype of a 666.7 kVA single-phase transformer with current limitation and recovery-under-load capabilities. The transformer represented a 2 MVA three-phase unit and was equipped with a superconducting primary and secondary winding and a conventional iron core, operated at 77.3 K.
In 2015, Callaghan Innovations presented results of a three-phase transformer project, for a fully superconducting transformer based on *Re*BCO-superconducting tapes, operated with sub-cooled liquid nitrogen at a temperature of 66 K [GSJA13, GSB+11, SGP+12]. The transformer had a three-phase complex power of 1 MVA and featured 12 mm wide *Roebel-cable* as a conductor in the secondary winding to ensure a perfect transposition of the single parallel tapes and hence, a homogeneous current distribution as well as reduced AC-losses. The transformer was also demonstrated in the power grid.
From the compilation presented in Table 1.2, it can be seen that all transformer featuring a fault current limitation are based on *Re*BCO-materials. The reason for this is that *Re*BCO-materials provide a significantly lower room-temperature resistance to critical current ration than BSCCO-materials [Kal11, Mel12]. This is due to the fact that the mechanical, electrical and thermal stabilization of BSCCO-conductor materials has to be made of silver, and therefore have a significantly reduced room-temperature resistance in comparison to superconducting *Re*BCO-tapes. This significantly reduces the effectiveness of BSCCO-conductor materials in fault current limiter applications, based on the resistance increase after the current-triggered quench of the superconductor.

2 Transformer Design Process

This chapter describes the design process which was used to construct and optimize a superconducting current limiting transformer with recovery-under-load.
Chapter 2.1 gives an overview of the single components of a superconducting current limiting transformer, e.g. the cryostat for enclosing the superconducting winding, the current leads and the winding former for mechanically suspending the fragile superconducting tape of the winding. This requires changes to the fundamental transformer design equations in order to augment the design process towards a superconducting current limiting transformer.
The design process starts with fundamental transformer design equations derived from the design processes of a normalconducting transformer. This part of the design process is described in Chapter 2.2.1. The fundamental transformer design equations are used to determine feasible designs for the transformer size, the amount of superconducting and normalconducting material, the iron core size and iron core weight as well as to give estimates for the transformer efficiency and the transformer short-circuit impedance.
The most important difference between the design process of a conventional transformer and a superconducting current limiting transformer is the prediction of the current limitation characteristics as it is presented in Chapter 2.2.2. For the design process shown in this work the current limitation prediction is done by numerical simulations. The goal of this simulation is to evaluate the limitation characteristics of a specific transformer design and the recover-under-load behavior after a fault current.
Intermediate results of Part I and Part II of the design procedure are shown in in Chapter 2.3.1 and in Chapter 2.3.2 respectively. The presented design process results in the transformer demonstrator that was constructed, manufactured and tested during the course of this work.
In Chapter 2.3.3 the detailed dimensioning and the parameters of the designed 1 MVA-class laboratory demonstrator are given.

2.1 Fundamental Transformer Design

The 1 MVA-class transformer demonstrator is designed as a single phase unit instead of a three phase design. The single phase setup allows for the proof of concept as well as for all required performance and current limitation tests with reducing the amount of materials for the construction and assembly to roughly one third, compared to a three phase transformer. The rated complex nominal power S_{n} of the single phase transformer was set to 577.4 kVA. This can be converted to an equivalent three-phase unit with a corresponding rated complex nominal power of 1.732 MVA.
The transformer design is not optimized for a long-term grid operation or for high efficiency but instead for effective current limitation during short test cycles in a laboratory environment.
For further simplification of the test setup, the transformer is equipped with a simple cryogenic system based on an open bath cryostat operated at ambient pressure. Despite of the simplicity of the cryogenic system it is still possible to perform a detailed loss analysis, as it is shown and explained in more detail in Chapter 5.1 and Chapter 6.3.
A schematic overview of the arrangement of the different transformer components is depicted in Figure 2.1. The following subsections give a more detailed overview of the specifications and the limitations for the conceptual transformer design.

Short-Circuit Impedance and Current Limitation

A superconducting current limiting transformer offers fail-safe intrinsic current limitation in case of a fault. For the demonstration of an efficient current limitation the presented transformer design should not exceed a short-circuit impedance value of 3 %. Otherwise the transformer impedance reduces the prospective short-circuit current, $I_{\mathrm{prosp.}}$ with inductive means, instead of being reduced by the temperature and resistance increase of the superconducting winding.
The fault duration and the consequent current limitation of the transformer are crucial parameters for the transformer design. The longer the current limitation lasts, the higher the temperature of the superconductor increases due to the Joule-heating in the material. For this transformer design the duration of the current limitation was therefore defined as a maximum of 6 half-waves (HW) or 60 ms at a grid frequency f_{grid} of 50 Hz. With modern switchgear it is easily possible to detect and disconnect a fault in a power grid in less than 60 ms in a reproducible and reliable manner [RJ89, Sla41].

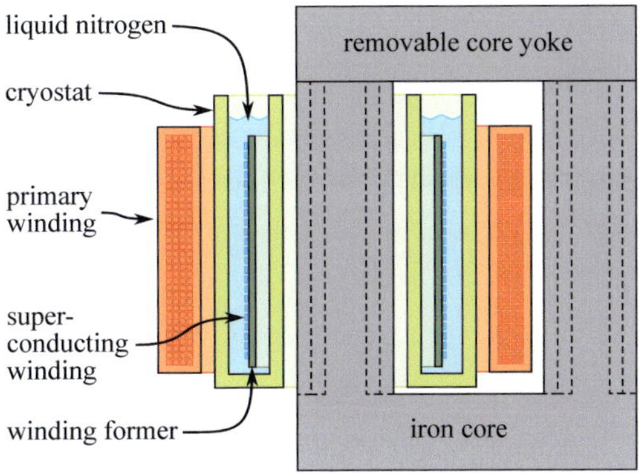

Figure 2.1: Schematic of conceptual transformer design and the arrangement of the main components (drawing is not to scale)

Transformer Windings

The main focus of the transformer design discussed in this thesis lies on an efficient current limitation rather than on a reduction in losses or the improvement of other properties. Thus, only one transformer winding is manufactured from superconducting material.

From an electrical and physical point of view, there is no difference between the fault current limitation being caused by the primary or by the secondary winding. However, the secondary transformer winding is usually operated at lower voltages, which makes the construction of the winding, including the electrical insulation more simple.

A voltage ratio U_p / U_s of 20 kV / 1 kV is chosen for the design. This ratio constitutes a common and realistic voltage ratio for a transformer of this power-class. The dimensions of the windings are kept to the same winding height to ensure the best possible magnetic coupling between the two windings and therefore the lowest possible values for the short-circuit impedance u_k.

For the composition of the winding itself, a stacked arrangement of superconducting tapes is used in order to investigate on the current distribution between the parallel conductors.

Superconducting Material

*Re*BCO-tapes are used as superconducting material for the secondary transformer winding. The *Re*BCO-conductor materials are described in Chapter 1.2. For reducing the AC-losses in the transformer winding, a 4 mm wide superconducting tape was chosen [BI93, Ber11, AKY$^+$04].

To manufacture a transformer winding, a winding former is required. This former keeps the superconducting tapes in position and mechanically supports and protects the tapes from the Lorentz-forces during nominal operation of the transformer and during a short-circuit limitation. This winding former contributes with its mechanical dimensions to the additional spacing that is required for the utilization of the superconducting material. Thus its dimensions are integrated into the transformer design procedure described in section 2.2.1.

An additional design criteria of the superconducting material is the maximum temperature T_{max}, that can be tolerated during a quench, without irreversible thermal degradation. Experiments show, that temperatures above 400 K to 600 K can lead to thermal degradation of the superconducting *Re*BCO-tapes [IAM$^+$11, LGD05, SKN08]. For this transformer project the maximum temperature T_{max} is decided to be set to 350 K. The maximum temperatures in the different transformer designs were estimated with the use of a transient simulation method as discussed in section 2.2.2.

Another important property of the superconducting material is the thermal and electrical stabilization thickness of the tape composition. This factor has a huge influence on the quench and current limitation characteristics of a superconducting transformer as well as on the Joule-heating in the superconducting tape. Therefore, the copper stabilization thickness is one of the main influencing parameters for the current limitation and the recovery-under-load characteristics of the transformer design.

Iron Core

There is a huge variety of widely used iron core designs for commercially available power transformers. For this project a U-I-core combination consisting of a U-shaped bottom iron core part and a removable iron core yoke (I-part) is used, as shown in Figure 2.1. The U-I-core is simple in its construction and common for a single phase experimental transformer. Of great importance for this project is the removable core yoke, since this allows easy access to the transformer windings.

The type of transformer core used in the assembly has minor influence on the core mass and the consequent iron core losses, the core dimensions and the winding arrangement. The transformer iron core is further described in Chapter 4.2.

Cryostat

For the transformer demonstrator, only the secondary winding, is made of superconducting material. This requires a cryostat that encloses the secondary superconducting winding only, while leaving the primary winding and the iron core at room-temperature operation conditions. For the proof-of-concept and for keeping the transformer design simple and robust, the cryogenic cooling of the superconducting secondary winding is done with liquid nitrogen at ambient pressure in an open cryostat. This results in a operating temperature for the superconducting material during nominal operation of 77.3 K. In addition to the cryostat itself, also the lid of the cryostat and the current leads, supplying current to the superconducting winding through the cryostat lid, have to be considered in the transformer design. The transformer cryostat is described in more detail in Chapter 4.4.

2.2 Transformer Design Equations

This chapter presents the whole design process as it was developed for the transformer that is manufactured and tested in this thesis. The design process can be separated into two parts:

- Part I: Basic design equations for a fundamental transformer design and
- Part II: Prediction of the current limitation behavior of the resulting superconducting transformer design

Figure 2.2 depicts the transformer design process in a flowchart showing the two different parts, the input parameters and the iteration loops.
Part I is based on an iterative design process using the enhanced basic transformer design equations. It is explained and discussed in Chapter 2.2.1. The input parameters for Part I of the design process are a set of predefined transformer and material parameters. Part I of the design process can be repeated in an iterative manner in order to optimize the basic transformer design. This is shown by the iterative loop I around Part I of the design process in Figure 2.2.
Part II in the design process is devoted to the current limitation characteristics and the estimation of the recovery-under-load behavior. It is described in detail in Chapter 2.2.2. Part II uses the transformer design parameters, as acquired in Part I of the design process, in combination with further predefined simulation and material parameters to generate a transient simulation model. This transient model can be applied for an iterative optimization of the current limitation and recovery-under-load properties, as indicated by the iterative loop II in Figure 2.2.
The design procedure further allows to iterate over both parts of the design process to allow the most effective optimization of the transformer designs towards specific desired characteristics. This full iterative loop is depicted as "iterative loop I + II" in Figure 2.2.

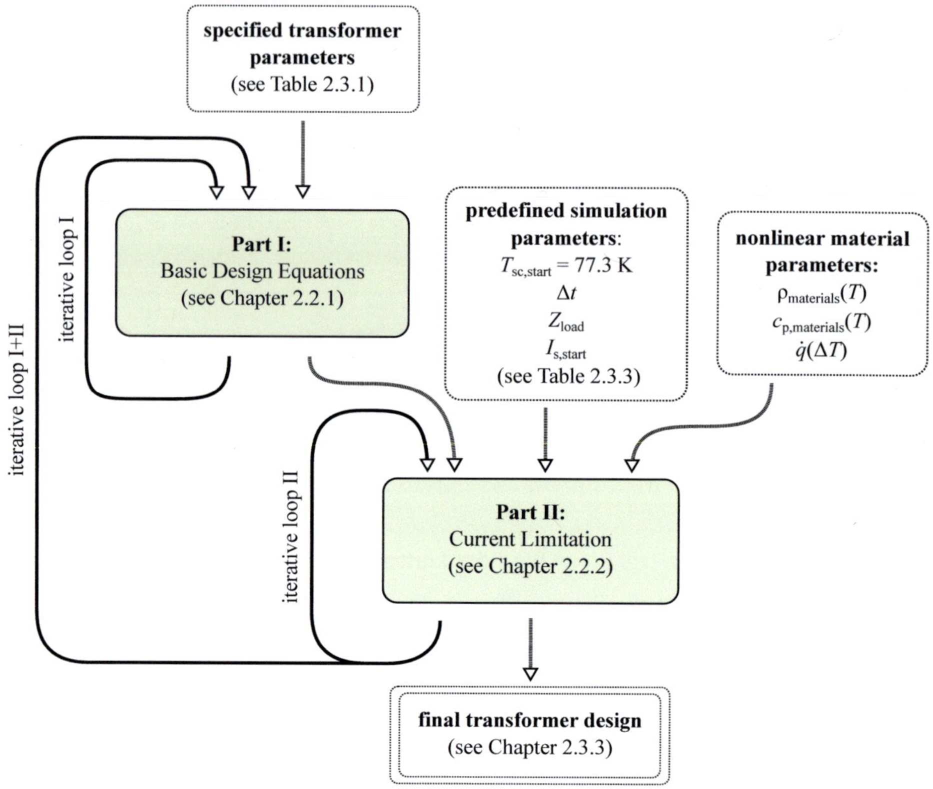

Figure 2.2: Flow chart illustrating the transformer design process. The whole process is here divided into two phases. Part I uses adjusted basic transformer design equations and Part II applies a transient simulation process to estimate the transformer current limitation behavior

2.2.1 Basic Design Equations

The fundamental design of a superconducting current limiting transformer is based on the design equations of a conventional transformer. These design equations are well established among transformer manufacturers and they are well documented in the literature [Fla93, Geo09, Har06, DPF+10]. The same fundamental design procedure can be used for superconducting transformer designs with and without current limiting properties.

In the fundamental transformer design process the usage of superconducting material was regarded in the layout, only by adding an estimated additional spacing in the design around the superconducting secondary winding. This additional spacing represents the room for the cryostat around the superconducting winding, the current leads for transporting the current into and out of the winding as well as a winding former and liquid

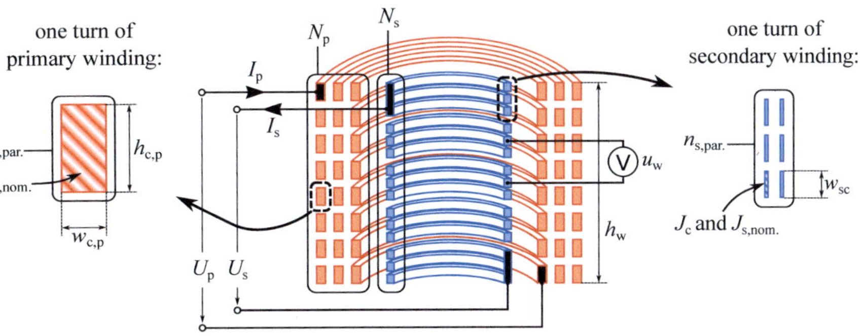

Figure 2.3: Symbols and nomenclature as used during part I and part II of the presented transformer design process (drawing is not to scale)

nitrogen reservoir inside the cryostat. This adjustment does influence the full transformer design, e.g. via a change in the necessary window size of the iron core, a different iron core mass, altered magnetic coupling between the windings and consequent changes in the short-circuit impedance.

The sketch shown in Figure 2.3 gives an overview of the symbols and the nomenclature used in the equations of the following chapters to describe the transformer design process. One of the main variables of a transformer design, is the turn-to-turn voltage u_w between the single turns in the transformer windings. With altering the turn-to-turn voltage of the transformer design, the dimensions of the windings change and the magnetic coupling as well as the iron core cross section area changes consequently. With adjusting the turn-to-turn voltage, the whole transformer design can be influenced and changed to desired values with respect to the iron core dimensions and weight, winding dimensions, transformer efficiency and the short-circuit impedance. It is common in a transformer design process to specify the turn-to-turn as a range of values, resulting in a range of designs with different properties.

Stepwise Calculation of Basic Transformer Design

As a first step of the fundamental design procedure the number of turns in the primary winding N_p and in the secondary winding N_s is calculated from the primary voltage U_p or the secondary voltage U_s and from the turn-to-turn voltage u_w, as described in Equation 2.1 and Equation 2.2.

$$N_p = \frac{U_p}{u_w} \tag{2.1}$$

$$N_s = \frac{U_s}{u_w} \tag{2.2}$$

In a next step the nominal turn ratio a can be calculated following Equation 2.3. The nominal primary and secondary current can be obtained from the transformer nominal complex power S_n in combination with the primary or secondary voltage. The nominal turn ratio is one of the parameters that is not dependent on the turn-to-turn voltage u_w, since the primary and the secondary voltage are within the predefined parameters.

$$\frac{U_p}{U_s} = \frac{I_s}{I_p} = \frac{N_p}{N_s} = a \tag{2.3}$$

The third step of the fundamental design process determines the cross section area A_{Fe}, for the main limb of the transformer iron core. A possible approach for this is via the law of induction.

$$u_s = -N_s \frac{d\Phi_h(t)}{dt} \tag{2.4}$$

The transformer main magnetic flux through the secondary winding in expressed as Φ_h. Its derivation multiplied with the number of turns in the secondary winding N_s yields the voltage induced in the secondary winding u_s as a result of the change in the magnetic flux Φ_h in the transformer iron core.

The magnetic flux Φ_h in Equation 2.4 can be replaced by the product of the magnetic flux density and the cross section area of the iron core.

$$\Phi_h(t) = B(t)\, A_{Fe,eff.} \tag{2.5}$$

For a sinusoidal change of the magnetic flux density in the transformer iron core, the law of induction from Equation 2.4 can be rewritten as shown in Equation 2.6.

$$U_p = N_p\, 2\pi\, f_{grid}\, A_{Fe,eff.} \frac{B_{max}}{\sqrt{2}} \tag{2.6}$$

If Equation 2.6 is solved for the effective cross section area of the iron core $A_{Fe,eff.}$, it leads to Equation 2.7 which allows the calculation of the total iron core cross section area A_{Fe} depending only on specified and on already determined parameters as well as on the turn-to-turn voltage u_w.

$$A_{Fe} = \frac{1}{\sqrt{2}\,\pi\, f_{grid}\, k_{Fe}\, B_{max}}\, u_w \tag{2.7}$$

In Equation 2.7, f_{grid} represents the grid frequency and k_{Fe} stands for the stacking factor of the laminated iron core. This factor specifies how much of the apparent (or total) cross section area of the iron core is actually available for the transportation of magnetic flux.

Due to the laminated structure of conventional iron cores and the electrical insulation between the sheets of laminated core material, this factor is smaller than 1 and typically around 0.96 in core materials for power transformers [DPF+10]. The maximum magnetic flux density in the iron core B_{max} is the predefined value for the iron core material. The maximum flux density needs to be kept underneath the saturation regime of the iron core material to eliminate hysteresis losses and to keep the eddy-current losses in the core sheets as low as possible.

The next step of the fundamental design process is the determination of the number of parallel conductors in the primary winding $n_{p,par.}$ and in the secondary winding respectively $n_{s,par.}$. They can be calculated from the nominal primary and secondary current, the specified conductor dimensions or superconducting tape width and the maximum current density for the conductor materials.

For the normal conducting primary winding this is shown in Equation 2.8. In this equation the height and width of the conventional conductor is specified as $h_{c,p}$ and $w_{c,p}$ and the assumed maximum current density $J_{p,nom.}$ is given in A/m^2.

$$n_{p,par.} = \frac{I_p}{h_{c,p}\, w_{c,p}\, J_{p,nom.}} \tag{2.8}$$

Equation 2.9 is used to determine the number of parallel conductors in the superconducting secondary winding. The maximum current density in the conductor material is specified via the critical current per width of the superconducting tapes J_c given in A/m together with a reduction factor for the critical current density β. The reduction factor is necessary in order to include the decrease in critical current caused by the magnetic field experienced by the single tapes in the superconducting winding. The width of a single superconductor is represented by w_{sc} whereas the height of the superconductor can be neglected here.

$$n_{s,par.} = \frac{I_s\, \sqrt{2}}{w_{sc}\, J_c\, \beta} \tag{2.9}$$

The magnetic field in a transformer is usually highly inhomogeneous and can only be precisely determined via a numerical simulation. Depending on the transformer parameters and on the current limitation characteristics, the reduction factor can be roughly estimated based on the analytically calculated magnetic field for the secondary winding. If necessary, this estimate can be corrected in a later step of the transformer design, based on numerically determined field distributions.

With the number of parallel conductors in the windings, with the conductor dimensions and with the required insulation distance between the conductors, the number of layers

in the winding can be calculated.
At this point it is useful to restrict the height of the windings and keep the windings to the same height. This improves the magnetic coupling between the windings. Further, the total length of conductors used in the design and the diameter of the central core limb can be calculated from the winding dimensions.
In the following step of the fundamental transformer design process the iron core dimensions can be calculated. This is done with the total iron core cross section area, the winding dimensions as well as with the electrical insulation distances. This step concludes the dimensioning of the transformer components.
The remaining design steps described in the following are necessary to calculate and estimate the short-circuit impedance and stationary performance characteristics of the resulting transformer design.

Calculation of Short-Circuit Impedance

In the next step of the transformer design process the magnetic stray field between the windings, $H_{\sigma,\mathrm{gap}}$ is calculated. Equation 2.10 is used for determining the magnetic stray field between the windings, depending on the number of turns in the primary winding N_p, the primary nominal current I_p and the height of the primary winding $h_{\mathrm{w,p}}$. It should be mentioned that Equation 2.10 does not provide a exact solution for the stray field but rather a valid approximation for a solenoid with a length several times larger than its radius [Whe28].

$$H_{\sigma,\mathrm{gap}} = \frac{N_\mathrm{p}\, I_\mathrm{p}}{h_{\mathrm{w,p}}} \tag{2.10}$$

The following step of the fundamental transformer design process is the calculation of the short-circuit impedance u_k. This can be done via the energy stored in the total magnetic stray field H_σ of the transformer. The total magnetic stray field is also an indication for the magnetic coupling between the two transformer windings.
Equation 2.11 shows the calculation of the energy stored in the stray field as a sum of three components, the stray field in the primary winding $H_{\sigma,\mathrm{p}}$, the stray field in the gap between the windings $H_{\sigma,\mathrm{gap}}$ and the stray field in the secondary transformer winding $H_{\sigma,\mathrm{s}}$.

$$W_\sigma = \iiint \frac{1}{2}\,\mu_0\, H_{\sigma,\mathrm{p}}^2\,\mathrm{d}v + \iiint \frac{1}{2}\,\mu_0\, H_{\sigma,\mathrm{gap}}^2\,\mathrm{d}v + \iiint \frac{1}{2}\,\mu_0\, H_{\sigma,\mathrm{s}}^2\,\mathrm{d}v \tag{2.11}$$

To avoid a complex algebraic or numerical solution for the precise field distribution in and between the windings, the assumptions for the magnetic field distributions, as shown

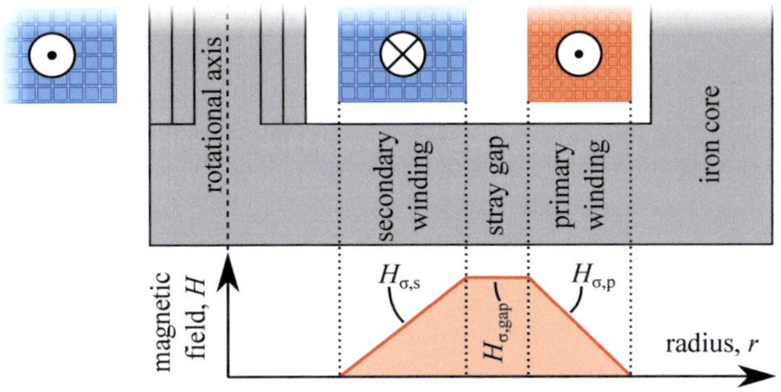

Figure 2.4: Simplified magnetic field distribution in the winding arrangement of the transformer in relation to the mechanical position of the windings and the iron core. Only a part of the total cross section view is shown (drawing is not to scale)

in Figure 2.4, are used. These are based on a homogeneous stray magnetic field between the windings, according to 2.10, and on linear gradients for the stray fields inside the windings.

From the energy stored in the magnetic stray field W_σ of the transformer windings, it is possible to calculate the stray inductance L_σ of the transformer design following Equation 2.12

$$L_\sigma = \frac{2\,W_\sigma}{I_\mathrm{p}^2} \tag{2.12}$$

Assuming the field gradients, as depicted in Figure 2.4, it is also possible to further simplify the calculation of the stray inductance, as shown in Equation 2.13. This direct approach uses only geometrical parameters from the transformer design as the thickness of the windings and the stray gap, th_p, th_s and th_gap, as well as the medium radius of the the windings and the stray gap, r_p, r_s and r_gap. Equation 2.13 however, can only be applied, if both transformer windings are of equal height h_w.

$$L_\sigma = \frac{2\pi\,\mu_0\,N_\mathrm{p}^2}{h_\mathrm{w}}\left(\frac{r_\mathrm{s}\,th_\mathrm{s}}{3} + r_\mathrm{gap}\,th_\mathrm{gap} + \frac{r_\mathrm{p}\,th_\mathrm{p}}{3}\right) \tag{2.13}$$

From the stray inductance L_σ the short-circuit impedance can be calculated following Equation 2.14. The input parameters here are the complex nominal power S_n, the grid frequency f_grid and the primary voltage U_p as well as the stray inductance L_σ.

$$u_\mathrm{k} = 100\% \cdot S_\mathrm{n}\,L_\sigma\,2\,\pi\,\frac{f_\mathrm{grid}}{U_\mathrm{p}^2} \tag{2.14}$$

The calculation of the transformer stray magnetic field, the short-circuit impedance as well as of the transformers stray inductance is further described in detail in the literature [Geo09].

Estimation of Transformer Losses

The total estimated transformer losses $P_{\text{loss,total}}$ can be calculated with Equation 2.15. The single components of the total losses in the transformer will be explained in the following. The dielectric losses, P_{c} in the insulation materials of the transformer assembly are neglected in this loss estimation.

$$P_{\text{loss,total}} = \frac{(P_{\text{loss,AC}} + P_{\text{loss,CL}} + P_{\text{loss,cryo}})}{\eta_{\text{cooling}}} + P_{\text{loss,Fe}} + P_{\text{loss,p}} \tag{2.15}$$

For the assessment of the AC-losses in the superconducting material, $P_{\text{loss,AC}}$ an estimation based on the approach for a single superconducting strip in self field is applied [Nor70]. The approach used by Norris is modified by dividing the complex geometry of the superconducting winding into its single winding turns. Then the average magnetic field, as experienced by each of these turns in the winding, was calculated using a finite element simulation model, using the calculated geometric parameters and the given current in the superconducting coil. The current distribution in the superconductor itself was assumed to be homogeneous across its width and not effected by the magnetic field. From the local magnetic field and the magnetic field angle, the reduction of the critical current density was calculated for each turn in the winding separately, based on measurement values for $J_{\text{c}}(B_{\text{ext.}}, \alpha)$, the critical current density depending on the magnetic flux density and the magnetic field angle. With this adjusted critical current density in every turn of the winding, the Norris approximation was used to estimate the losses in the single turns, treating them as single, separate conductors in self field. The sum of the calculated losses over all turns in the winding is then assumed to be a estimate for the AC-loss part of the transformer efficiency model $P_{\text{loss,AC}}$.
The loss components for the thermal energy conducted along the current leads into the cryogenic environment $P_{\text{loss,CL}}$ and the heat conducted through the cryostat walls $P_{\text{loss,cryo}}$ are estimated based on empiric values for the cryostat losses and literature values for the heat intake along the the optimized current leads into the cryogenic environment [Wil83, CL12, See98].
Since the AC-losses, the cryostat losses and the thermal losses through the current leads are all affecting the amount of evaporated coolant in the transformer setup, they have to be divided by the efficiency of the cryogenic refrigeration unit η_{cooling}, as shown in Equation 2.15.

In addition to the losses in the cryogenic environment, the estimation of the total losses also includes the magnetization and eddy-current losses in the iron core material $P_{\text{loss,Fe}}$.

$$P_{\text{loss,Fe}} = m_{\text{Fe}}\, P_{\text{Fe,spec.}} \tag{2.16}$$

These are calculated according to Equation 2.16, from the iron core mass m_{Fe} and from the specific iron core losses $P_{\text{Fe,spec.}}$ given in W/kg. The specific iron core losses are typically specified by the manufacturer of the iron core material and depend on the maximum magnetic flux density B_{max}.
The Joule-heating in the primary winding is calculated with Equation 2.17. This is calculated from the resistivity of the conductor material $\rho_{\text{cond.}}$ in the primary winding, the single conductor length l_{p} and dimensions $h_{\text{c,p}}$ and $w_{\text{c,p}}$, the number of parallel conductors $n_{\text{p,par.}}$ and the nominal current in the winding I_{p}.

$$P_{\text{loss,p}} = \frac{I_{\text{p}}^2\, \rho_{\text{cond.}}\, l_{\text{p}}}{h_{\text{c,p}}\, w_{\text{c,p}}\, n_{\text{p,par.}}} \tag{2.17}$$

Refinement of Fundamental Transformer Design

In order to refine the resulting transformer designs, additional parameters need to be specified and considered in the design process. These additional parameters are for example the specification of the cryostat, the cryostat lid and the current leads into the cryogenic environment. Further refined parameters have to be specified for the electrical insulation distances in the normalconducting and the superconducting winding. Also, the exact design and the dimensions of the iron core and engineering features and limitations for the general assembly of the transformer need to be determined in order to generate a more detailed and complete transformer design.
Another factor in the refinement of the design process is the availability of the components. The medium voltage normalconducting windings are only manufactured in certain winding dimension, a certain insulation distance and winding height. This has to be regarded in the design process and other transformer parameters have to be adjusted in order to keep the desired properties for the resulting transformer design. The same restrictions apply for the iron core and for the dimensions of raw materials, as they are utilized in the construction of the cryostat or the superconducting winding former.
Since the overall design changes with adjusting single parameters, it is often necessary to repeat the whole design process in several iterations until a satisfying result is achieved while considering possible limitations given by materials and the availability of components.

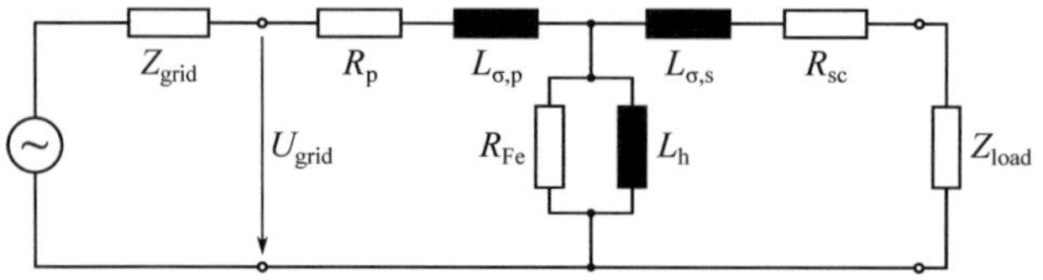

Figure 2.5: The circuit diagram shows the equivalent circuit of the transformer on which the transient simulation of the transformer behavior is based on

2.2.2 Current Limitation Characteristics

The next step of the design procedure is the simulation of the temperature increase in the superconducting winding and of the current limitation characteristics for the transformer designs, yielded by the basic design equations of Part I. This is done with a transient numerical simulation method that is applied to each of the designs yielded by Part I of design process. The simulation is based on a zero-dimensional transformer model, neglecting any inhomogeneous behavior along the superconducting tapes in the winding and any geometry dependent properties. This assumption can be made, because the transition from the superconducting state into the normalconducting regime is fast and because the fault current in the secondary winding $I_{fault,s}$, during the limitation of the fault, is significantly larger than the nominal secondary current I_s.

Figure 2.5 shows the equivalent circuit for the superconducting transformer demonstrator. It is identical to the equivalent circuit of a conventional transformer, except for the resistance of the secondary winding, which is here replaced by the resistance of the superconductor in the secondary winding R_{sc}. The transformer main inductance L_h and the resistive component R_{Fe} , representing the magnetization losses in the iron core, are considered not relevant for the current limitation characteristics and are therefore neglected in the simulation. Further, the two parts of the transformer stray inductance, $L_{\sigma,p}$ and $L_{\sigma,s}$ are combined in the total stray inductance L_σ as calculated with Equation 2.13.

For the numerical simulation the *forward-Euler method* is used, with the temperature of the superconducting tape T_{sc} and the secondary transformer current I_s being the dependent variables throughout the simulation. The forward-Euler method is a first-order numerical concept for solving ordinary differential equations.

The flow chart shown in Figure 2.6 depicts the transient simulation loop for obtaining the temperature of the superconducting winding and the current limitation for a specific set of parameters. The boxes in the top part of the figure list the input parameters for the simulation loop. The parameters listed in box (1) on the left side are results from Part I of the design process.

In addition to the adopted parameters it is necessary to specify boundary conditions for

the simulation and simulation specific parameters as shown in box (2). These are the initial temperature of the superconducting tape $T_{\text{sc,start}}$, the initial current in the secondary winding $I_{\text{s,start}}$, the duration of the single simulation time steps Δt and the transient value for the transformer load Z_{load}.
Specified in input box (3) are the non-linear material parameters for the superconducting tape model, such as the specific heat $c_{\text{p,materials}}(T)$ and the resistivity $\rho_{\text{materials}}(T)$ of the superconducting tape materials as well as the heat flux density into the liquid nitrogen bath $\dot{q}(\Delta T)$. These parameters depend on the temperature of the superconducting tape and are based on literature values [Eki06, Roy10, MC64].
The first step after the initiation of the transient simulation loop is the calculation of the resistance of the superconducting winding R_{sc} as shown in box (A) of Figure 2.6 with Equation 2.18. This is done separately for the different layers of materials in the tape composition. The resistance values of the single material layers are treated as parallel electric resistors to calculate the total resistance of the superconducting winding. The temperature value for determining the resistivity is the temperature from the last simulation step $T_{\text{sc}}(n-1)$ or, the initial temperature $T_{\text{sc,start}}$ for the first iteration.
For temperatures below the critical temperature T_{c}, the resistivity of the *Re*BCO-layer in the tape is calculated according to the power-law, depending on the transported current in this layer. For temperatures above the critical temperature the resistivity of the *Re*BCO-layer is calculated based on the temperature dependent literature values for normalconducting *Re*BCO-material. Those two models, the power-law and the resistive model, are treated as a parallel circuit to calculate the resulting current and temperature dependent resistivity of the *Re*BCO-material. This model is documented in the literature [Gri04] and often referred to as the *bounded power-law*.
In the second step, shown in box (B), the total impedance of the full equivalent transformer circuit, $Z_{\text{tot.}}$ is calculated with Equation 2.19 from its different in-series components. Here Z_{grid} is the internal impedance of the grid, R_{p} is the resistance of the primary winding, X_{σ} is the stray reactance of the transformer, R_{sc} represents the resistance of the superconducting winding converted to the primary side and Z_{load} is the transformer load. The load is converted to the primary side as well and can be switched from nominal load to a specific short-circuit load during a simulated fault event.
The calculation of the current in the secondary winding of the transformer is the next step, shown in box (C) of Figure 2.6 with Equation 2.20. It is obtained from the grid voltage $U_{\text{grid}}(n-1)$, the resistive components R_{p}, R_{sc} and R_{load}, the current in the previous step of the simulation $I_{\text{s}}(n-1)$, the duration of a simulation time step Δt as well as from the the stray inductance L_{σ} and the inductive component of the transformer load L_{load}.

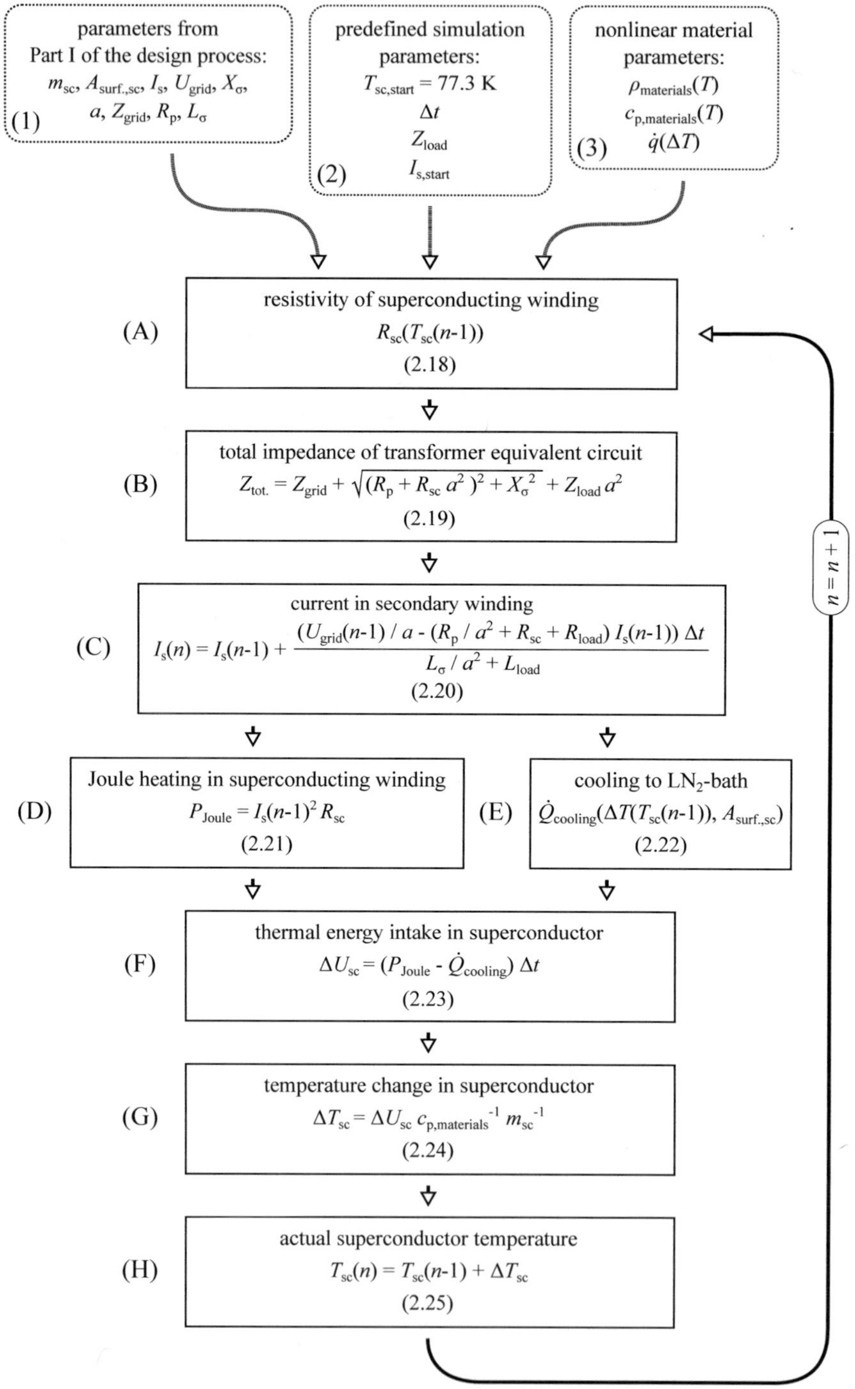

Figure 2.6: The depicted flow chart illustrates the forward-Euler approach for the simulation of the transient current limitation of a transformer design

After acquiring the updated current in the secondary winding with Equation 2.20, the heat intake into the winding can be calculated via the Joule-heating P_{Joule} from the current $I_s(n-1)$ and the already obtained resistance of the superconducting winding R_{sc}, as seen in box (D) with Equation 2.21. In parallel to the heat intake, the outward heat flux into the nitrogen bath, $\dot{Q}_{\mathrm{cooling}}$ can be calculated with using Equation 2.22 in box (E). This is done with the temperature of the superconducting tape from the last simulation step $T_{sc}(n-1)$ and the surface area of the superconductor $A_{\mathrm{surf.,sc}}$. The superconductor surface area $A_{\mathrm{surf.,sc}}$ describes the surface of the superconductor, which is directly adjacent to the liquid nitrogen and therefore available for heat transfer. In a stacked conductor arrangement, this only applies for the outer surface of the stack of superconducting tapes. The total thermal energy balance in the superconductor of the secondary winding can be obtained with Equation 2.23 as shown in box (F) of the flow chart in Figure 2.6.

As shown in box (G), the temperature change in the conductor material $\Delta T_{sc}$is calculated with Equation 2.24 from the change in the thermal energy ΔU_{sc} in the superconductor, the specific heat capacity of the different materials in the superconducting tape $c_{\mathrm{p,materials}}$ and the mass of the superconducting tape in the secondary winding m_{sc}. The total heat capacity is obtained from the temperature dependent literature values for each layer separately and added to the total heat capacity of the superconducting tape composition.

In the last step of the transient simulation loop shown in box (H), the temperature of the superconducting material in the secondary winding is updated with the change in the temperature ΔT_{sc} and the superconductor temperature from the last time step $T_{sc}(n-1)$, as shown with Equation 2.25.

Since the error per step of the applied Euler-method is proportional to the square of the step size, it is necessary do prove the independence of the results from the size of the time steps Δt in the simulation to ensure the plausibility of the results.

2.3 1 MVA-Class Transformer Design

In this section the transformer design equations, as introduced during part I in Chapter 2.2.1 and during part II in Chapter 2.2.2, are used to generate a set of feasible transformer designs for a 1 MVA-class single phase transformer demonstrator.

The transformer designs resulting from part I of the design process are based on a set of predefined parameters. They vary dependent on the turn-to-turn voltage and are shown in Table 2.3.1 together with intermediate design results.

Chapter 2.3.2 shows the application of the transient simulation introduced in Chapter 2.2.2 as part II of the transformer design process. The transient simulation is used to de-

termine the current limitation characteristics of the set of transformer designs as acquired from part I of the design process.
Chapter 2.3 concludes in Chapter 2.3.3 with the presentation of a selected transformer design based on the specified parameters. The chosen design fulfills the requirements as specified and proves to exhibit efficient current limitation behavior. The transformer manufactured and tested through the further course of this thesis is based on the presented transformer design.

Table 2.1: List of parameters specified for part I of the presented transformer design procedure

parameter	symbol	design value
primary voltage	U_{p}	20 kV
secondary voltage	U_{s}	1 kV
complex nominal power	S_{n}	577.4 kVA
grid frequency	f_{grid}	50 Hz
short-circuit impedance	u_{k}	< 3 %
maximum flux density in iron core	B_{max}	1.6 T
iron core stacking factor	k_{Fe}	0.965
iron core material density	ρ_{Fe}	7650 kg / m^3
cross section area of normal conducting primary conductor	$A_{\mathrm{p}} = h_{\mathrm{c,p}} \cdot w_{\mathrm{c,p}}$	6.3 mm · 5 mm
cross section area of superconducting secondary conductor	$A_{\mathrm{s}} = h_{\mathrm{sc}} \cdot w_{\mathrm{sc}}$	0.3 mm · 4 mm
maximum nominal current density, primary	$J_{\mathrm{p,nom.}}$	1.2 A / mm^2
maximum nominal current density, secondary	$J_{\mathrm{s,nom.}}$	160 A per 4 mm tape width
critical current correction factor	β	0.65 (65 % of original J_{c})
estimated electrical insulation distances	-	2 mm around conductors, 20 mm around primary winding
additional spacing for cryostat, winding former and LN_2-reservoir	-	300 mm top spacing, 100 mm bottom spacing, 45 mm left / right

2.3.1 Basic Design Equations

Table 2.1 lists all parameters which are specified prior to the design process. The parameters include the nominal operating voltages, U_{p} and U_{s} of the transformer and the grid frequency f_{grid} as well as of the complex nominal power S_n. Also listed are specific transformer design parameters like the iron core stacking factor k_{Fe}, the iron core material density ρ_{Fe}, the cross section area of the conductors and estimates for additional spacing around the windings to regard for electrical and thermal insulation.

The table further contains the design limits of the transformer and of the materials used, e.g. the maximum short-circuit impedance, the maximum magnetic flux density in the core material B_{max} as well as the maximum current densities in the windings, $J_{p,nom.}$ and $J_{s,nom.}$.

As mentioned in Chapter 2.2.1, the following design process is carried out for a range of parameters sets at different turn-to-turn voltages. This in turn leads to a set of transformer designs, all based on identical parameters as listed in 2.1 but with a different turn-to-turn voltage. For the design process, the turn-to-turn voltage range was set to be between 0 V and 100 V per turn in the transformer windings.

Figure 2.7 shows the change of the total cross section area A_{Fe} of the transformer core versus the turn-to-turn voltage of the windings. As it can be expected from Equation 2.7, the total cross section area increases proportional to the turn-to-turn voltage since all other parameters in the equation are given values.

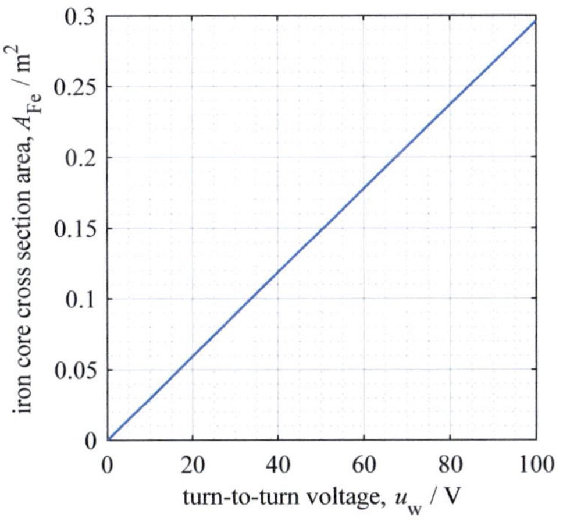

Figure 2.7: Total iron core cross section area versus the turn-to-turn voltage u_w of the transformer

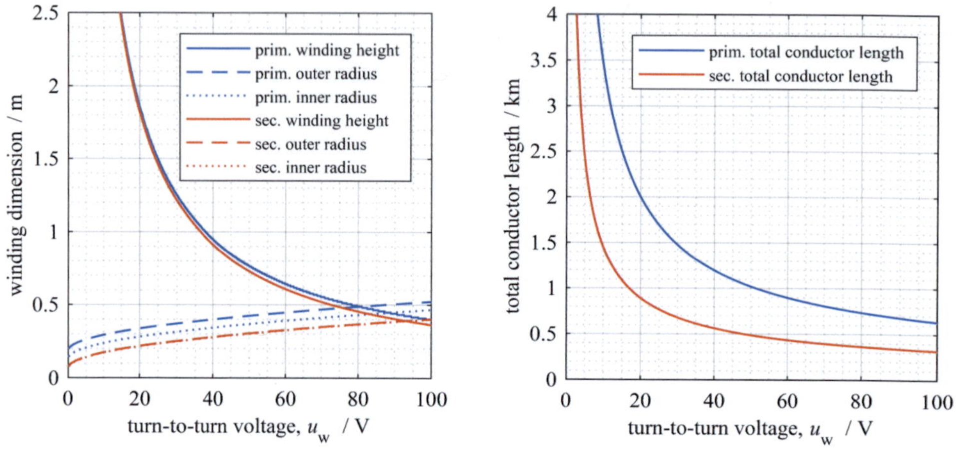

Figure 2.8: Winding dimensions versus turn-to-turn voltage u_w (left) and resulting total length of conductors in windings versus turn-to-turn voltage u_w of the transformer windings (right)

The calculated dimensions of the windings as well as the required total length of conductor material can be seen in Figure 2.8. These results can be acquired from the total iron core cross section area A_{Fe} in combination with the specified electrical and thermal insulation distances listed in Table 2.1. The total conductor length can then be calculated from the winding dimensions and from the specified conductor material, using Equation 2.8 and Equation 2.9.

For the presented design process a U-I-iron core design was chosen for the calculations on the iron core. Figure 2.9 shows the dependency of the iron core dimensions and the iron core mass on the turn-to-turn voltage. The iron core dimensions (not including the iron core cross section area A_{Fe}) are calculated during part I of the design process from the dimensions of the windings and the additional specified spacing.

The short-circuit impedance u_k versus the turn-to-turn voltage, as calculated during part I of the design process, is shown in Figure 2.10. It decreases with higher values for the turn-to-turn voltage because the dominant term in Equation 2.13 is the square of the number of turns in the primary winding. Additionally shown in Figure 2.10 is the maximum specified short-circuit impedance of 3 %. It can be seen that only transformer designs with a u_w larger than 40 V fulfill this specification and are therefore possible design candidates.

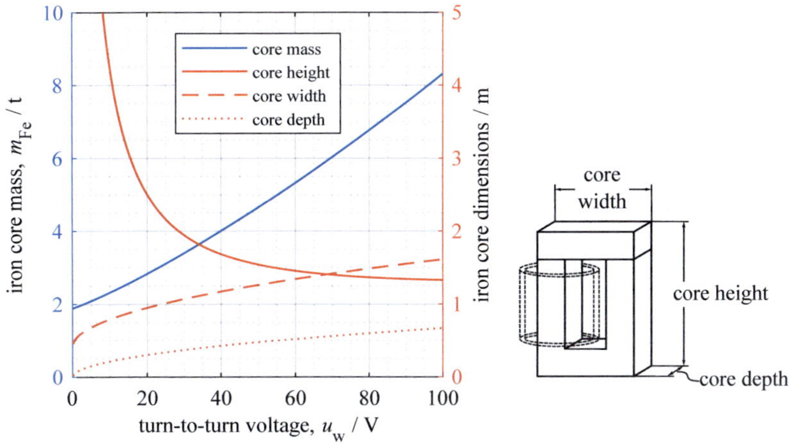

Figure 2.9: Iron core dimensions and resulting iron core mass versus the turn-to-turn voltage u_w of the transformer. The additional sketch explains the denotation of the iron core dimensions

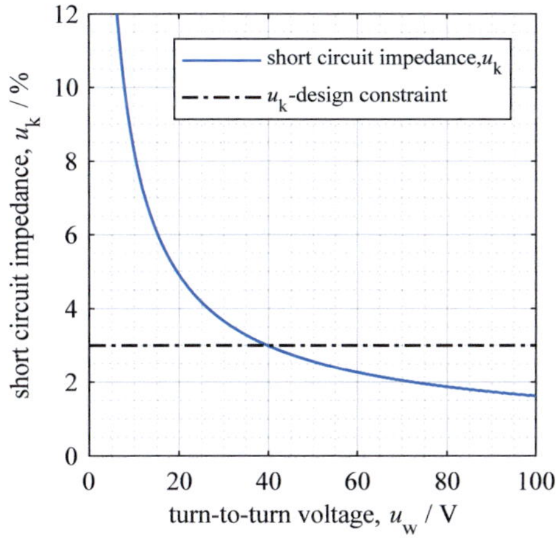

Figure 2.10: The graph visualizes the short-circuit impedance u_k versus the turn-to-turn voltage u_w together with the design limit for u_k. Only a turn-to-turn voltage higher than 40 V yields a short-circuit impedance below the specified maximum of 3 %

Table 2.2: Additional parameters assumed for the estimation of the transformer losses and the total transformer efficiency η_{total}

parameter	symbol	design value
estimated AC-losses of superconducting material	$P_{loss,AC}$	see Chapter 2.2.1
heat intake through the current leads [CL12, Wil83]	$P_{loss,CL}$	50 W / kA
heat intake through the cryostat walls	$P_{loss,cryo}$	10 W / m^2
efficiency of the cryogenic system	$\eta_{cooling}$	0.065 (6.5 %)
specific iron core losses (manufacturer)	$P_{Fe,spec.}$	1.01 W / kg
resistivity of the conductor material in the primary winding	$\rho_{cond.}$	temperature dependent literature value for conductor material [Eki06]

Estimation of Transformer Losses

In addition to the constructional and electrical parameters yielded by part I of the design process, as shown in Figure 2.7 to Figure 2.10, it is also possible to estimate the transformer losses. Table 2.2 list the additional parameters specified for the estimation of the transformer losses.

The loss calculation during part I of the transformer design process is carried out as described in the last part of Chapter 2.2.1 with Equation 2.15, 2.16 and 2.17. The left graph in Figure 2.11 shows the different loss contributions of the transformer versus the turn-to-turn voltage. In the graph, the losses affecting the cryogenic part of the setup have been divided by the efficiency of the refrigeration system $\eta_{cooling}$, according to Equation 2.15.

From the acquired nominal load losses it is then possible to calculate the transformer efficiency at nominal load for the different resulting transformer designs versus the turn-to-turn voltage. The efficiency at nominal load is presented in the right graph of Figure 2.11. The graph shows, that the maximum efficiency of a transformer design based on the specified parameters is achieved for a turn-to-turn voltage of around 24 V.

Further Design Adjustments

The primary winding of the transformer design was determined to an electrical height of 905 mm. This was done for availability reasons for commercially available resin-cast transformer windings.

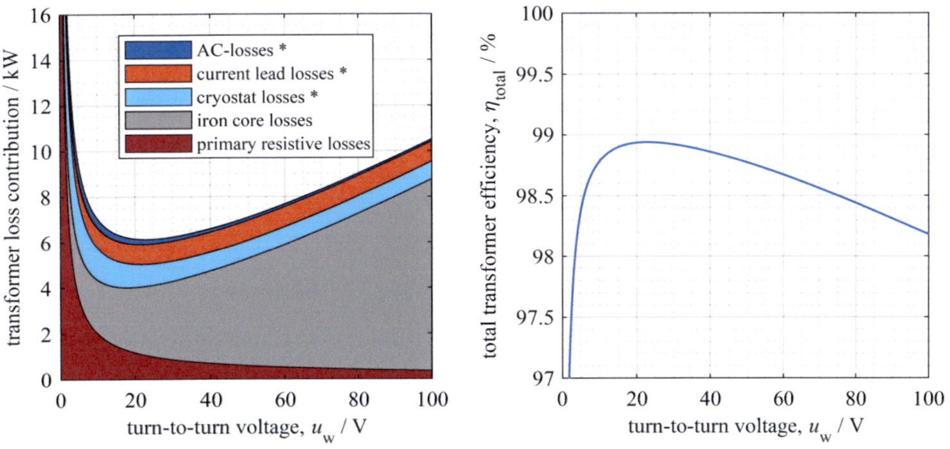

Figure 2.11: The left graph shows the different loss contributions of the transformer designs versus the turn-to-turn voltage. The asterisk (*) in the legend indicates that the losses have been divided by the efficiency of the cooling system. Plotted in the right graph is the resulting total transformer efficiency η_{total}. It decreases significantly for lower values of the turn-to-turn voltage u_{w} and reaches a maximum at 24 V

Due to this aspect, the transformer designs evaluated in the following chapter are not fully unbound, but rather assume a fixed winding height, $h_{\text{w,p}}$ for the primary and $h_{\text{w,s}}$ secondary winding, versus the variable turn-to-turn voltage u_{w}. This consequently results in a increasing spacing between the conductors in the secondary superconducting winding, for higher values of u_{w}, or in multiple layers of superconductors composing the secondary winding, for lower turn-to-turn voltages u_{w}.

2.3.2 Current Limitation Characteristics

In the following section the transient evaluation of a set of transformer designs, as acquired from part I of the design process in Chapter 2.3.1, is presented. For this, the transient evaluation process for the prediction of the current limitation characteristics, as described in Chapter 2.2.2, is applied.

In order to conduct the simulation, further parameters need to be specified to generate the transient simulation model. Table 2.3 lists the parameters that are additionally defined. Specified in the table are transient parameters such as the nominal and fault load, $Z_{\text{load,nom.}}$ and $Z_{\text{load,fault}}$, as well as the duration of the fault t_{fault}. To define the material composition of the superconducting tape in the transient model, the specific heat of its components needs to be calculated from the thickness of the different layers, given in Table 2.3. The table further lists the length of the simulation time step Δt, that proved to be sufficiently

Table 2.3: List of parameters used for generating the transient simulation model and to simulate the current limitation and recovery-under-load characteristics of the superconducting transformer

parameter	symbol	design value
length of simulation time step	Δt	10 μs
nominal transformer load (here only real part)	$Z_{\text{load,nom.}}$	1.73 Ω
load during fault (here only real part)	$Z_{\text{load,fault}}$	0.0253 Ω
grid impedance	Z_{grid}	3.054 Ω
fault duration	t_{fault}	60 ms (six half-waves)
maximum allowed superconductor temperature	T_{max}	350 K
critical temperature of superconducting tape	T_c	90.4 K
critical current of superconducting tape	J_{sc}	160 A (per 4 mm wide tape)
copper stabilization thickness	th_{Cu}	110 μm
substrate thickness	th_{HY}	50 μm
silver cap layer thickness	th_{Ag}	2 μm

small to guarantee step size independent results. With the specified parameters the transient simulation is then executed for each turn-to-turn voltage dependent transformer design given by part I of the design process.

The graphs in Figure 2.12 present results from the transient simulation for one transformer design with a turn-to-turn voltage of 40 V.

The left graph shows the transformer load Z_{load} (blue line), as it is applied to the transformer model during the transient simulation. The load changes from the transformer nominal load to a low value during the simulated fault on the secondary side of the transformer and then back to the nominal load, to simulate the clearance of the fault from the power grid. The same graph also shows the increase of the simulated temperature of the superconducting tape in the current limiting transformer winding.

The graph on the right side of Figure 2.12 depicts the result of the simulation - the limited current (blue line) of the transformer design that is caused by the quench and subsequent temperature increase of the superconducting material in the secondary winding. The graph also shows the prospective short-circuit current $I_{\text{prosp.}}$ (black dashed line), as it would be expected, if the transformer would not be equipped with a current limiting superconducting winding. The calculation of the prospective current is done for all transformer designs to allow a comparison of the efficiency of the current limitation in

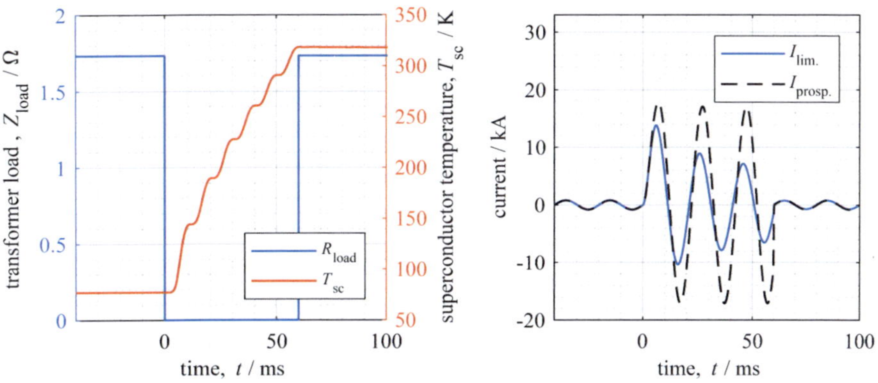

Figure 2.12: Example for a transient simulation result for the simulated transformer load and the temperature increase (left) and the limited transformer current as well as the prospective current (right). The graphs show the numerical simulation for a single transformer design based on parameters yielded by part I of the design process at a turn-to-turn voltage of 40 V

the different designs via the ratio between the limited current and prospective current, $I_{lim.}$ divided by $I_{prosp.}$.

The prospective current is calculated in the transient simulation process by assuming a superconducting tape with an infinite critical current density or, in other words, a resistance of the secondary winding R_{sc} that is zero. This method for calculating the prospective current is still considering the full inductive contribution to the current limitation of the transformer as well as the resistive contribution of the primary winding. The resulting prospective fault current is close to the fault current as it would be for an identical conventional transformer.

The results of the transient analysis are discussed in the following.

The maximum superconductor temperature after the limitation T_{max} is shown in the top graph of Figure 2.13. The visualization depicts the temperature of the superconducting winding versus the turn-to-turn voltage u_w as well as the given design constraint of 350 K as a maximum allowed conductor temperature, as defined in Table 2.3. It can be seen from the graph that designs with a turn-to-turn voltage above 45 V result in a maximum temperature after the limitation higher than the specified maximum of 350 K. Thus these parameter sets are not considered as possible transformer designs.

It can further be observed from the same graph, that the superconductor temperature only surpasses the critical temperature ($T_c = 90.4$ K) for u_w-values higher than 25 V. The reason for this is that the thermal mass of the superconducting winding is higher for lower turn-to-turn voltages. At the same time, the short-circuit current through the transformer is already significantly reduced by inductive means, caused by higher values

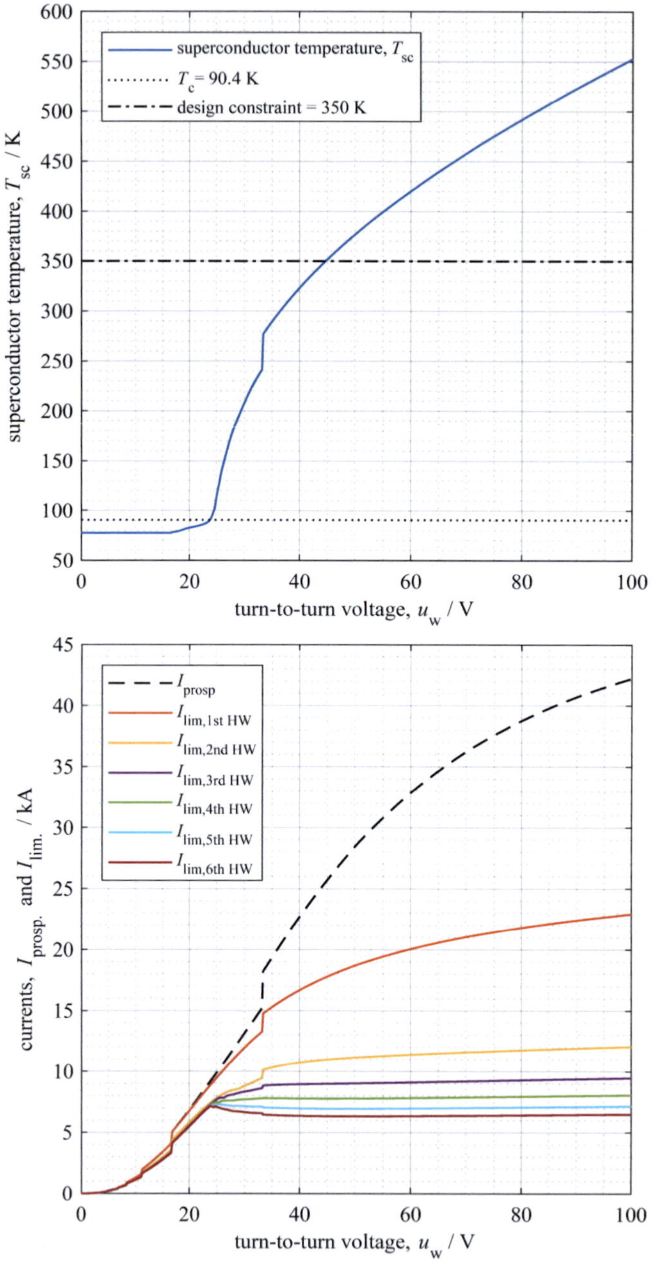

Figure 2.13: The top graph shows the maximum temperature T_{max} of the superconducting material after the six half-waves of fault current versus the turn-to-turn voltage u_w. Depicted in the lower graph are the maximum values of the currents in the peaks of each of the the six half-waves of a fault, versus the turn-to-turn voltage. The expected prospective fault current is represented by the black dashed line

for the short-circuit impedance u_k, as seen in Figure 2.10.
Due to this correlation, a significant current limitation by quenching the superconducting winding can only be expected for u_w-values above 25 V.
The visible step in the temperature curve, around a turn-to-turn voltage of 33 V, is caused by the change in the number of conductor layers in the secondary superconducting winding - for lower turn-to-turn voltage values the winding is composed of two or more layers, while there is only one layer in the secondary winding for values larger than 33 V. This is due to the fact that the height of the windings in the transformer design are fixed to 905 mm, as described in Chapter 2.3.1.
The prospective current $I_{prosp.}$ versus the turn-to-turn voltage u_w together with the expected limited current $I_{lim.}$ is shown in the bottom graph of Figure 2.13. The limited current in this graph is shown as the peak current values of the six simulated half-waves during a fault duration of 60 ms. The graph shows that the prospective current of the transformer designs increases significantly for higher values of the turn-to-tun voltage. The main reason for this, is the lower number of turns in the windings of the transformer. This reduces the resistance of the windings as well as the winding height. A smaller winding size in return, increases the magnetic coupling of the windings through a reduction of the stray field and therefore results in higher prospective short-circuit currents.
Similar to the top graph of Figure 2.13, the bottom graph as well shows a step in the expected currents for the design changing to a single layered superconducting secondary winding around a turn-to-turn voltage of 33 V. It can be seen, that a significant current limitation, especially in the first half-wave of the limited current, can only be expected for single layered transformer windings with turn-to-turn voltages above 33 V. This design experiences a higher temperature increase in the quenched superconductor material, what in turn results in a more efficient current limitation during the fault.
While showing the expected limited currents versus the turn-to-turn voltage of the different transformer designs, the bottom graph of Figure 2.13 does neither give any information about the reason for the current limitation of a specific design nor does it provide any indication about its composition.

This can be achieved with a visualization as shown in Figure 2.14. The graphs presented in this figure visualize the different current limiting components of the transformers equivalent circuit varying with the turn-to-turn voltage. The current limitation is composed by the following contributions:

- resistance of the superconducting secondary winding, R_{sc} (real component only, plotted in blue)
- resistance of the transformer short-circuit load, $Z_{\mathrm{load,fault}}$ (real component only, plotted in red)
- transformer stray reactance, X_{σ} (imaginary component only, plotted in yellow)
- the resistance of the primary winding, R_{p} (real component only, plotted in purple)
- impedance of the power grid, Z_{grid} (real and imaginary component, plotted in green)

The separate components and their variation via the turn-to-turn voltage either calculated from the parameters as results from part I of the transformer design procedure ($Z_{\mathrm{load,fault}}$, X_{σ}, R_{p}) and from part II, the transient evaluation of the transformer design process (R_{sc}). The power grid impedance is constant and solely depending on the the grid. For a better comparison of the values, the resistance of the superconducting winding and the transformer short-circuit load are converted to the primary side of the transformer.
The two graphs in the top row of Figure 2.14 show the composition of the current limiting factors in the transformer design as absolute values for the impedance of the equivalent circuit. Again, the left graph shows the composition versus the turn-to-turn voltage in the peak of the first half-wave of the limitation whereas the graph on the right depicts the current limiting factors in the sixth half-wave of the fault. The sum of the components used to calculate the prospective current for the different transformer designs is marked with a dashed black and white line in both graphs at the top of the figure. For smaller turn-to-turn voltages this value increases significantly. This is caused by the higher number of turns for the resistive part of the secondary winding and by the transformers stray reactance, which grows with the square of the turn number for decreasing values of u_{w}. Since the current limiting components aside the superconducting contribution are not experiencing transient changes during the current limitation, the parts underneath the dashed black and white line remain unchanged in the comparison between the first and the sixth half-wave.
The graphs in the bottom row of Figure 2.14 show the same current limiting factors but as a percentage share of all elements in the equivalent circuit instead. The left graph depicts

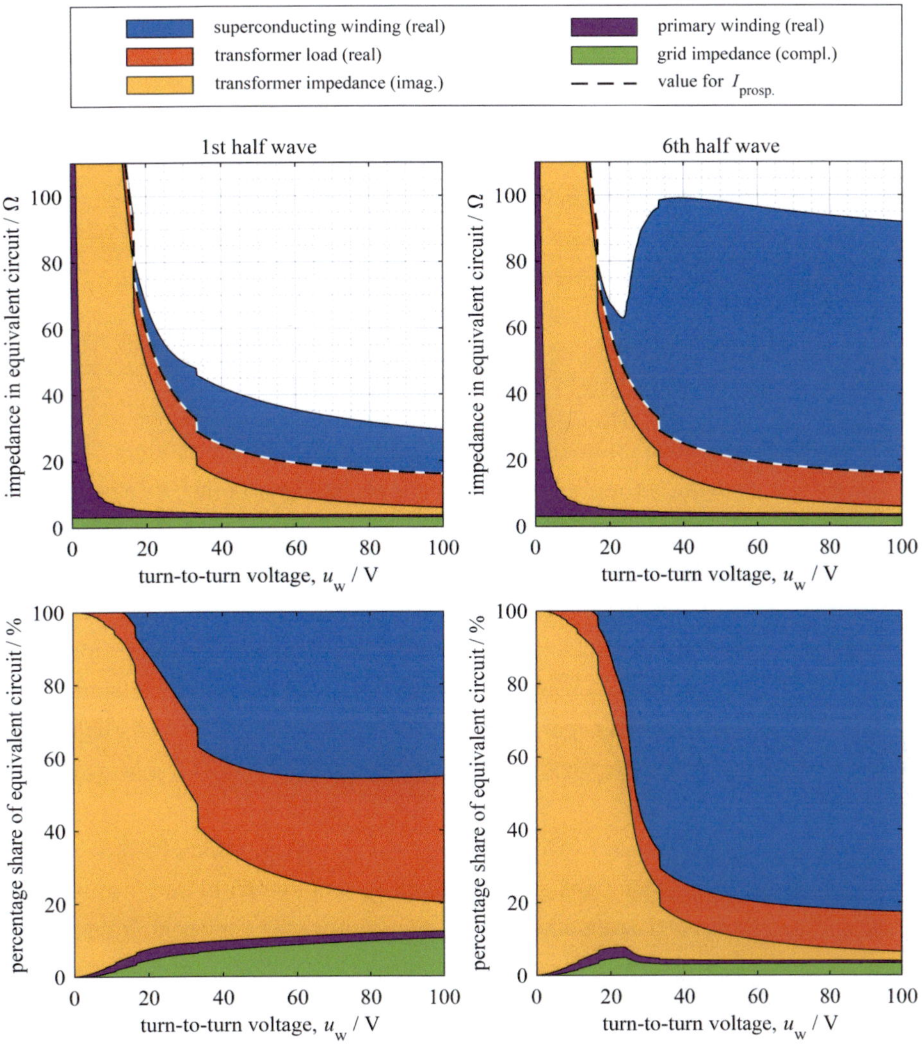

Figure 2.14: This graph visualizes the percentage share of the current limitation exhibited by the main components of the whole test circuit in the peak of the first half-wave of the current limitation versus the turn-to-turn voltage u_w. Assumed for this transformer design is a copper stabilization of the superconducting tape of $th_{Cu} = 110\ \mu m$

the composition versus the turn-to-turn voltage in the first half-wave of the short-circuit event, and the right graph shows the composition in the last half-wave respectively.
It can be seen that the influence on the current limitation due to the superconducting material in the secondary winding first appears for turn-to-turn voltages just above 15 V and increases up to values of u_{w} around 40 V. For higher values the percentage share on the current limitation remain almost stable for the first half-wave and increases only slightly for the limitation in the sixth half-wave.
The increase of the conductor temperature for higher u_{w}-values would suggest a higher specific resistance and therefore an increase of the current limitation. However this temperature increase for higher turn-to-turn voltages is also accompanied by a reduction in the total length of conductor material in the secondary winding and therefore compensates the effect of a improved current limitation.

Variation of Additional Design Parameters

During the transient design evaluation process shown so far, the transformer designs are only varied via changing the turn-to-turn voltage. However, another design factor for influencing the performance and operational characteristics of a superconducting current limiting transformer is the thickness of the copper stabilization layer in the superconducting tape. While being of negligible influence during the normal operation and the superconducting state of the transformer, the copper stabilization layer has a significant effect on the quench and current limitation characteristics of the superconducting winding in the transformer assembly. For this reason it is of importance to evaluate different possible transformer design depending on the copper stabilization of the utilized superconducting tape.
The following graphs depict different performance and current limitation characteristics of the same set of transformer design parameters as acquired from part I of the transformer design process and evaluated with the transient model shown in Figure 2.6. In contrary to the results presented previously, in this evaluation the designs are not solely varied over the turn-to-turn voltage u_{w} but also over the thickness of the copper stabilization th_{Cu}. The results of this evaluation are visualized in two-dimensional arrays with the investigated values plotted versus the copper stabilization thickness and versus the turn-to-turn voltage.
The previously shown graphs in Figure 2.13 and in Figure 2.14 are part of the acquired values presented in the following and are therefore represented by black and white dotted lines in the two-dimensional visualizations.
Figure 2.15 shows the maximum temperature T_{max} of the superconducting winding af-

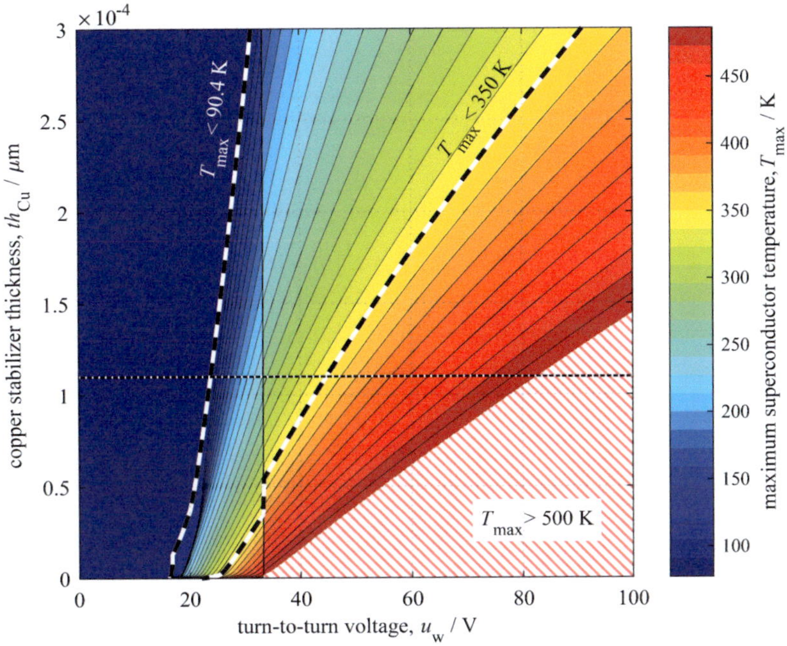

Figure 2.15: This graph shows the maximum temperature of the superconductor after the current limitation versus the turn-to-turn voltage (x-axis) and the copper stabilization thickness (y-axis). For better resolution the graph is limited to values for T_{max} < 500 K

ter a 60 ms fault, as simulated during the transient process. The temperature is shown depending on the thickness of the copper stabilization th_{Cu} (y-axis) and the turn-to-turn voltage u_w (x-axis). For the visualization, the results are limited to parameter combinations yielding a maximum temperature of 500 K or below, to increase the color resolution in the lower, reasonable temperature range.

The area in the graph with the maximum temperature T_{max}, not exceeding the critical temperature T_c of the superconductor and the area with a temperature lower than the specified maximum of 350 K, are indicated by dashed black and white lines. The area between the two dashed lines represents parameter combinations between the turn-to-turn voltage, u_w and the thickness of the copper stabilization, th_{Cu} which result in current limitation due to the heating in the superconducting material, while at the same time not exceeding the design limit for the maximum temperature after the limitation of 350 K.

It can be seen that the maximum temperature for transformer designs with higher stabilized tapes generally result in lower maximum temperatures over the whole range of turn-to-turn voltage values. This is caused by the higher thermal mass of the tapes and therefore a lower temperature increase during the Joule-heating process following the quench of the tape.

Figure 2.15 also shows the design change from two layers in the secondary winding to a single layered winding at a turn-to-turn voltage value of 33 V, as described for Figure 2.13 and Figure 2.14. The change to a single layer secondary winding creates a discontinuity of values that results in a vertical step of accumulated contour lines in the two dimensional plot. This discontinuity between a two and a one-layer design is apparent as well for the graphs shown in Figure 2.16 and Figure 2.17.

The two graphs shown in Figure 2.16 depict the contribution of the superconducting winding in the current limitation as the percentage share of the equivalent circuit, in which the top graphs shown the contribution in the first and the bottom graph shows the contribution in the sixth half-wave of the simulated fault. For a better comparison between the first and the sixth half-wave, both graphs were given the same range in the color scale.

It can be seen that the highest contribution in the first half-wave is accomplished for superconducting tapes with a maximum of 10 µm copper stabilizer. These tapes experience the most significant temperature increase, already in the first half-wave, and therefore provide the highest resistive contribution to the equivalent circuit. In the sixth half-wave, in contrast, the contribution to the current limitation is in the same range along the line, indicating the maximum temperature limit of 350 K. This shows the dependency of the current limitation on the conductor temperature.

Figure 2.17 presents the simulated current limitation of the whole transformer, including inductive and normal conducting components, in dependency of the copper stabilization thickness th_{Cu} and the turn-to-turn voltage u_{w}. Again, the current limitation in the first half-wave is shown in the top graph whereas the sixth half-wave is plotted in the graph at the bottom of Figure 2.17 and both graphs have the same color scale range.

The graphs show that the limitation in the first half-wave still depends, for the same turn-to-turn voltage, more on the copper stabilization than it does in the sixth half-wave. This is because the superconducting tapes with more copper stabilizer do not reach the higher and therefore more effective conductor temperatures already in the first half-wave. In the sixth half-wave of the fault, the current limitation flattens out for a wide range of combinations between the copper stabilizer thickness and the turn-to-turn voltage. This is shown in the bottom graph of Figure 2.17 with a wide area of of current limitations in the range between 15 % and 30 % of the prospective current.

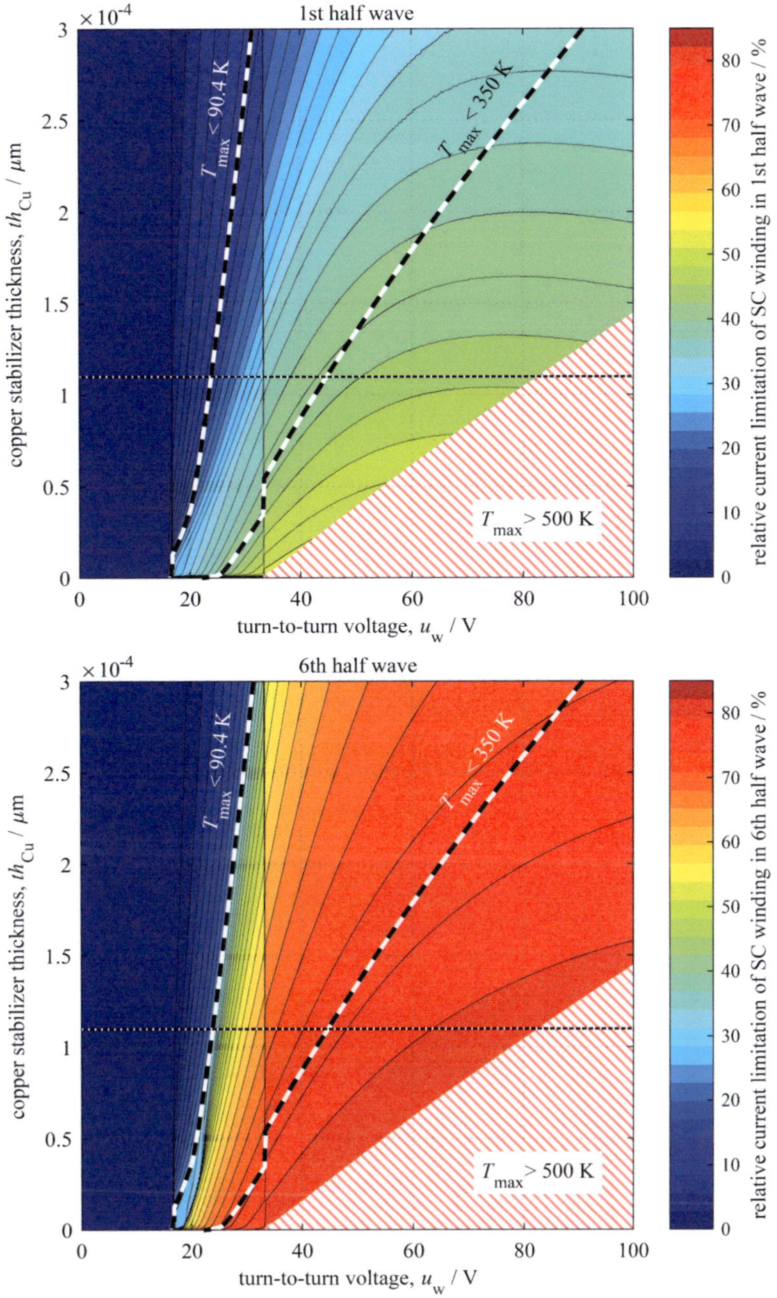

Figure 2.16: The graphs show the contribution of the superconducting winding in the current limitation as the percentage share of the equivalent circuit versus the turn-to-turn voltage (x-axis) and the copper stabilization thickness (y-axis). It is visualized for the first halve wave (top graph) and the sixth half-wave (bottom graph). For better resolution both the graphs are limited to values for $T_{max} < 500$ K

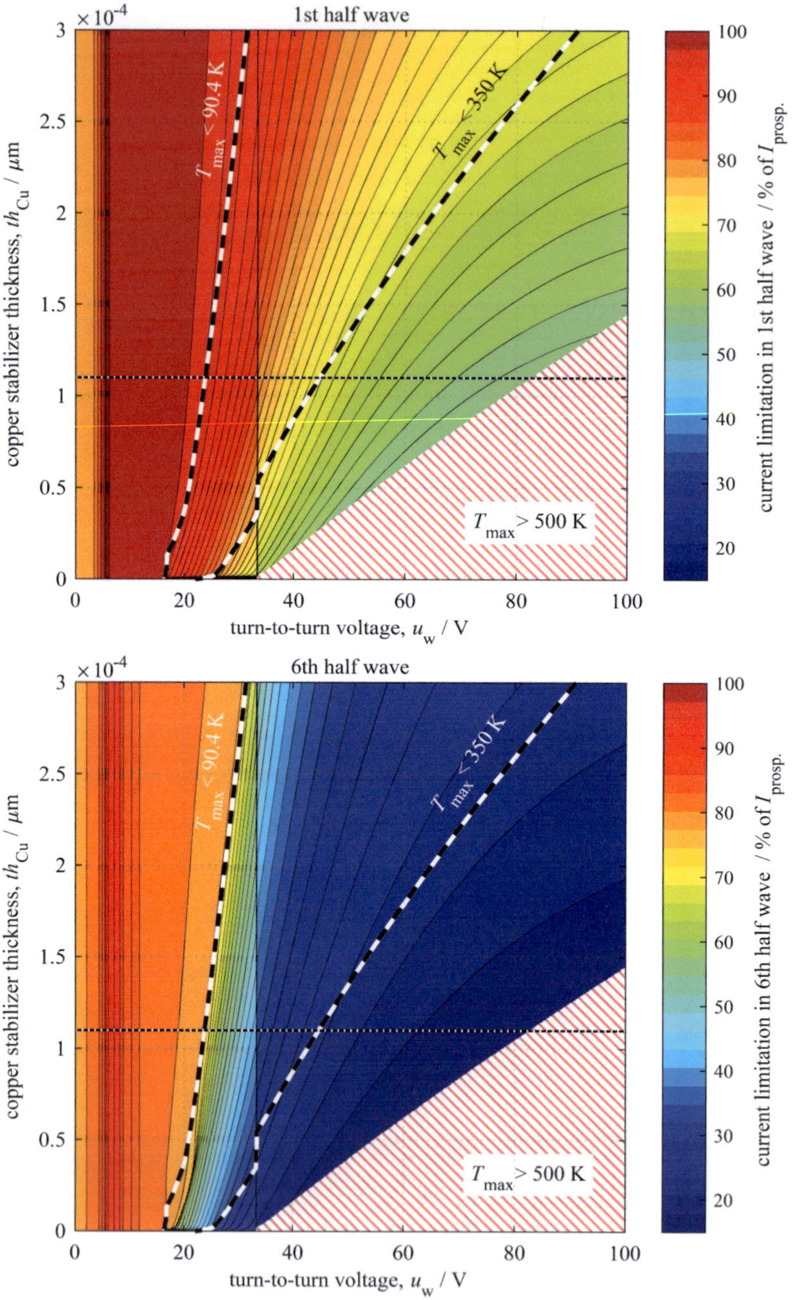

Figure 2.17: The graphs of this figure show the limited fault current as a percentage of the prospective current versus the turn-to-turn voltage (x-axis) and the copper stabilization thickness (y-axis). It is visualized for the first halve wave (top graph) and the sixth half-wave (bottom graph). For better resolution both the graphs are limited to values for T_{max} < 500 K

2.3.3 Transformer Dimensions and Properties

With the transformer design method, consisting of part I with the basic transformer design equations and the transient limitation evaluation in part II, different design parameter combinations were assessed.

The whole transformer design phase, as described in Chapter 2.2.1 and Chapter 2.2.2, is a iterative process, which requires the change and the adjustment of parameters in order to find a feasible transformer design that satisfies given performance requirements.

During the design process the iterative adjustment of parameters, especially the turn-to-turn voltage and the copper stabilization, was repeated until one feasible design constellation was found, which fulfills all given specifications, as for example the maximum allowed short-circuit impedance u_k and maximum temperature T_{max} after the limitation. This was done with the help of the visualizations as presented in Figure 2.15, Figure 2.16 and Figure 2.17, which allowed direct evaluation of different possible combinations between the turn-to-turn voltage u_w and the copper stabilization thickness th_{Cu}, under consideration of component availability and predicted transformer performance.

For the single phase 1 MVA-class transformer demonstrator a turn-to-turn voltage of 40 V, in combination with a copper stabilizer thickness of 110 µm was chosen. The design satisfies the requirements given and explained in Chapter 2.1 and listed in Table 2.1. The design utilizes only commercially available components for the conventional transformer primary winding and iron core. The transformer cryostat is custom build, but limited in the dimensions to the availability of tubular raw materials of the desired diameters. For the superconducting tapes, a commercially available superconductor is used with the given thickness of 110 µm of copper stabilizer and no further customization to the material.

With the specification of the turn-to-turn voltage u_w and the thickness of the copper stabilization th_{Cu} the electric transformer parameters are fully defined and the remaining parameters can be calculated. The additionally calculated parameters are listed in Table 2.4.

The different contributions to the total transformer losses are listed in Table 2.5. The losses are calculated for the specified turn-to-turn voltage of 40 V and a efficiency of the refrigeration system, $\eta_{cooling}$ of 6.5 %, as listed in Table 2.2.

Table 2.4: Main transformer design specifications and results of the design process consisting of part I and part II

	parameter	specified / resulting	design value
part I	turn-to-turn voltage	specified	40 V
	parallel conductor configuration	resulting	6 x stack of 2 tapes
	number of parallel tapes	resulting	12
	winding length	resulting	46.81 m
	total superconductor length	resulting	561.72 m
	transformer core mass	resulting	4001.3 kg
	turns in primary winding	resulting	500
	turns in secondary windings	resulting	25
	total iron core cross section area	resulting	0.1185 m^2
	short-circuit impedance	resulting	2.986 %
	maximum nominal magnetic flux density in stray gap	resulting	27.2 mT
part II	copper stabilizer thickness	specified	110 μm
	maximum superconductor temperature after limitation	resulting	323.2 K
	current limitation in 1st HW	resulting	16.67 kA (73.5 %)
	current limitation in 6th HW	resulting	6.35 kA (28.0 %)
	prospective current 1st HW	resulting	22.67 kA (100 %)

Table 2.5: Composition of transformer losses for the specified turn-to-turn voltage of 40 V. The values in parenthesis are not taking the efficiency of the refrigeration system, η_{cooling} of 6.5 % into account

contribution	value
AC-losses (without η_{cooling})	146.2 W (9.5 W)
current lead losses (without η_{cooling})	887.7 W (57.7 W)
cryostat losses (without η_{cooling})	827.7 W (53.8 W)
iron core losses	4041.3 W
primary winding losses	697.1 W
total transformer losses	6600.0 W

Figure 2.18 presents a sketch showing the main dimensions in the assembly of the resulting transformer design as a outcome of the design process:

Iron Core and Main Parameters - (A) and (B)

Section (A) of the transformer sketch shows the iron core dimensions, the placement of the windings and the cryostat around the rotation axis of the transformer. Shown in (B) are the dimensions of the cross section area of the stepped transformer core limbs. For the top and the bottom yoke a rectangular cross section area is used for simplicity reasons.

Transformer Cryostat - (C)

In section (C) of the sketch shown in Figure 2.18, the dimensions of the cryostat, the cryostat walls and the placement of the winding inside the cryostat are depicted. The placement of the secondary winding inside the cryostat includes additional room for the current leads into the cryogenic environment as well as for busbars and taps to connect the superconducting secondary winding.

Secondary Winding and Conductor Arrangement - (D) and (E)

Section (D) shows the conductor arrangement in the composition of the superconducting secondary winding. It can be seen that the twelve parallel superconducting tapes in the secondary winding are arranged in six parallel stacks, composed by two tapes each. This conductor configuration was specified in order to test the use of stacked conductors in current limiting windings. Section (E) of the sketch depicts a closeup of one of the conductor stacks showing the double layered insulation between the single tapes in one stack of the winding.

The transformer design results acquired with part I and part II of the design process throughout Chapter 2.2.1 and Chapter 2.2.2 give all basic parameters that are necessary for the manufacturing of the transformer while not containing any constructional details. If during the following manufacturing and assembly process any changes or improvements of the existing design are required for constructional reasons, it can be necessary to repeat the design iterations to ensure the continuous fulfillment for the required performance of the resulting transformer design.

A detailed description of the manufacturing of the transformer components and the constructional details of the full assembly will be given in Chapter 4.

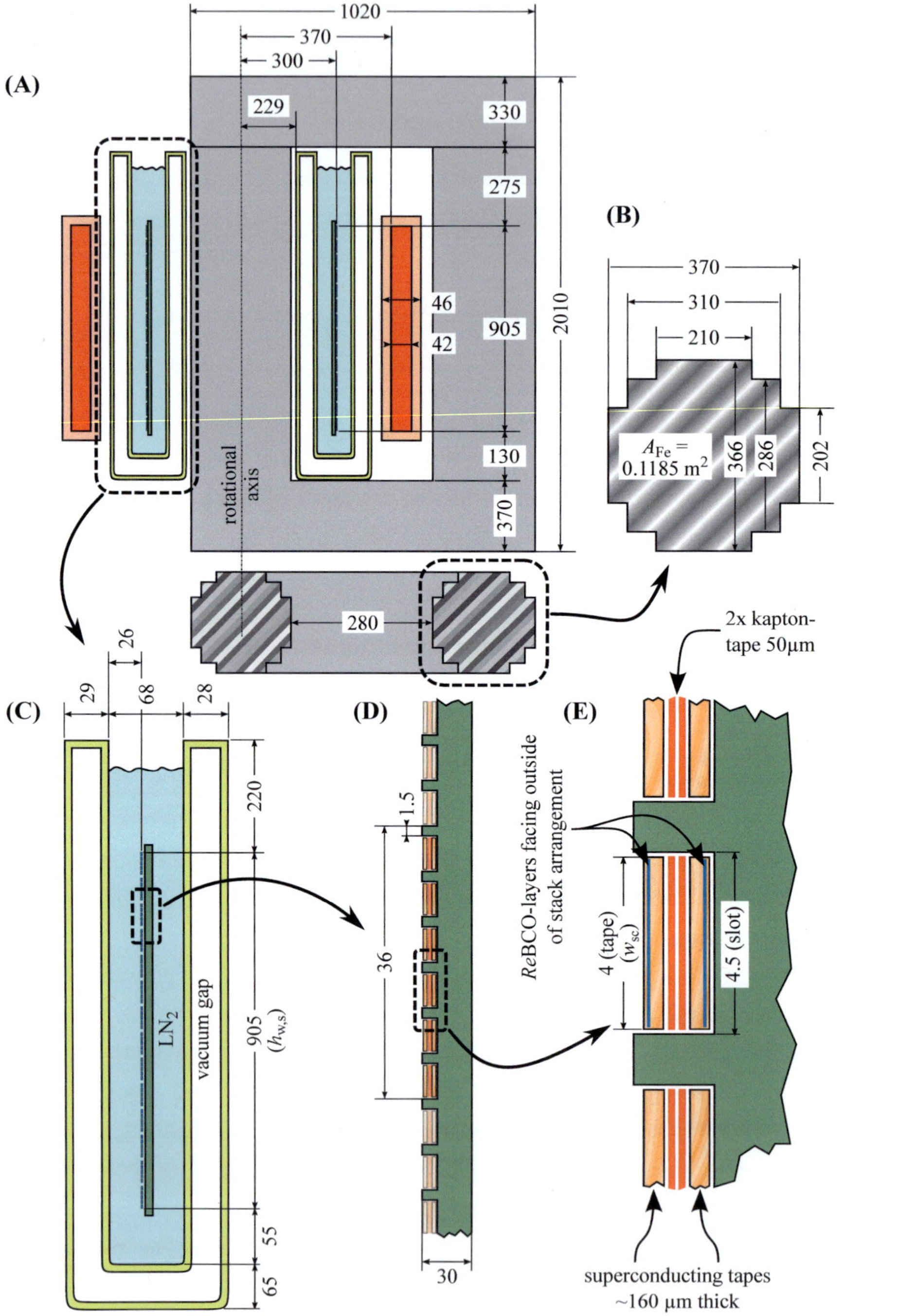

Figure 2.18: Final transformer dimensions after the iterative design process via the use of the fundamental transformer design process in combination with a transient performance prediction. All dimensions are given in millimeters (drawing is not to scale)

3 Preliminary Experimental Investigations

This chapter describes the experiments and investigations carried out in order to understand the properties and the behavior of the materials and techniques used for the design and the manufacturing of the superconducting current limiting transformer. The preliminary investigations comprise the characterization of the superconducting materials, as well as the efficiency and the general properties of the cooling inside an open bath of liquid nitrogen.

In order to select a commercially available superconducting tape for the application in a superconducting fault current limiting transformer, experiments were conducted on the quench behavior of different superconducting tapes as well as on the recovery-under-load behavior of those tapes. The quench characteristic of a superconducting tape is mainly determining the current limiting properties of a transformer.

Further investigations were performed to study the cooling mechanisms of the superconducting tapes inside a liquid nitrogen bath. The cooling is of great importance for a superconducting current limiter, especially if recovery-under-load is one of the desired operating features.

Furthermore this chapter proposes possible improvements on the heat exchange between superconducting materials and liquid coolants.

The preliminary experimental investigations were conducted in order to find a commercially available tape that satisfies all criteria required for the transformer design as it is depicted in Chapter 2.

3.1 Heat Transfer from Solids to Liquid Coolants

The heat flux $\dot{q}$ is the amount of heat energy transferred from the volume of a solid body through its surface and into another system with a lower temperature T. This type of heat transfer is also referred to as *cooling* and is usually expressed in W/cm^2. If the adjacent thermal system with lower temperature is a liquid coolant, then the heat exchange or cooling process is a complex combination between heat conduction, convective ther-

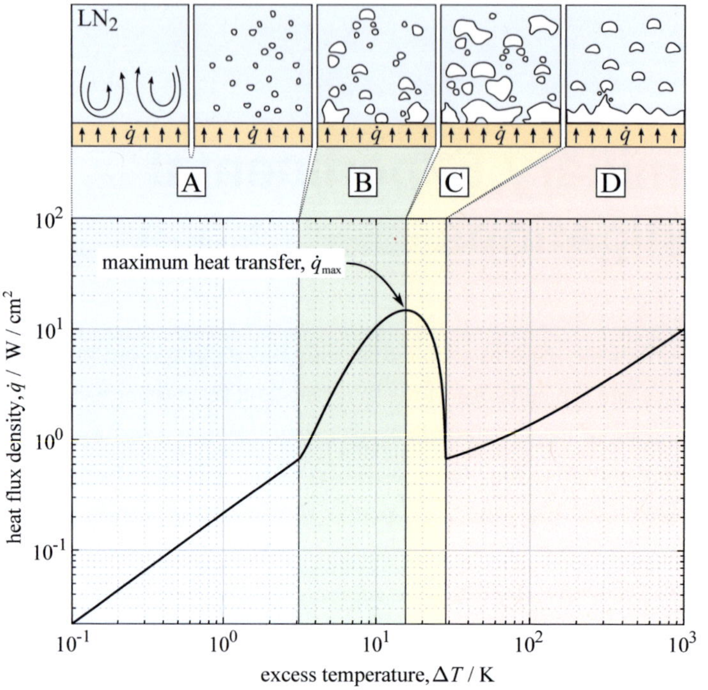

Figure 3.1: This graph shows the boiling curve for LN_2 at ambient pressure (T = 77.3 K) based on measurement values originally published by H. Merte et. al. [MC64]

mal energy transport, thermal radiation and cooling via phase transition (vaporization) [Kav01]. Therefore, the heat flux cannot be expressed as a constant value but varies depending on several factors. The excess temperature ΔT has the most significant influence on the efficiency of the heat transfer. This is the temperature difference between the surface of the solid and the temperature of the liquid coolant.

The heat transfer function for a solid super-heated body immersed in liquid nitrogen is depicted in Figure 3.1. This visualization shows the heat flux density as a function of the excess temperature. It is also referred to as the *boiling curve*. The heat transfer from the solid into the liquid coolant varies over several magnitudes depending on the excess temperature ΔT. This is due to the different phases of heat transfer that are active for different temperature differences between the surface of the solid and the coolant.

As shown in Figure 3.1 the heat transfer is divided into four regimes (A to D) with different heat transfer mechanisms.

Regime *A* shows the *natural convection boiling*. This heat transfer mechanism consists of mainly convection cooling as a heat transfer function and results typically in values for $\dot{q}$ that are well below 1 W/cm^2. For very low temperature differences in this regime there is no formation of bubbles at the interface between the solid and the liquid nitrogen and hence no visual evidence of a heat transfer taking place. For higher temperature differences (closer to regime B) small bubbles start to form at the surface of the solid and rise in the bath of coolant but often re-condense immediately. The higher amount of bubbles, forming with an increase of the excess temperature, improves the heat transfer to the coolant, resulting in a linear rise for $\dot{q}$ as depicted in the double-logarithmic plot of Figure 3.1.

With increasing excess temperature, the heat transfer function enters the regime of the so-called *nucleate boiling*, this is depicted as regime B in Figure 3.1. In the nucleate boiling regime the amount as well as the size of the bubbles increase and the bubbles do not re-condense anymore. This is due to the higher amount of liquid coolant being vaporized at the interface between both systems. In regime B the heat flux increases significantly with higher excess temperatures. One reason for this is the efficient cooling by vaporization (phase transition) of liquid nitrogen. The second factor that enhances the heat transfer in the regime of nucleate boiling is the higher number and bigger size of bubbles in the coolant. This in turn accelerates the convective part of the cooling by creating a stronger flow in the coolant allowing more liquid nitrogen being transported across the cooling interface. In regime B the heat flux increases by more than one order of magnitude and peaks around a value of 13 W/cm^2 for an excess temperature around 14 K. This peak in the heat transfer function is also referred to as the *maximum heat transfer* $\dot{q}_{\mathrm{max}}$.

If the excess temperature surpasses the point of the maximum heat transfer, the amount of vaporized liquid coolant increases further, meaning that more and more surface area of the solid exposed to the liquid coolant is covered with bubbles and not in direct contact with the liquid phase of the coolant anymore. This in turn does not allow for efficient heat transfer between solid and coolant and therefore reduces the values for the heat flux $\dot{q}$ significantly. This regime of the boiling curve in Figure 3.1 is labeled as C and usually referred to as *transition boiling*.

The fact that in regime C the excess temperature rises in combination with a descend in the heat transfer leads to an unstable thermal condition. This means, that if ΔT is increased to values above 14 K, less heat can be transferred to the coolant and assuming a constant heat source in the solid, the surface temperature of the solid increases, what in turn leads to a further increase of ΔT. This makes the exposed surfaces of the solid being further covered by a gaseous layer of evaporated coolant. In this condition, an interface between a solid and a liquid coolant will rapidly increase in temperature and therefore shift into regime D, the *film boiling regime.*

In the film boiling regime the entire surface of the solid is covered with a gaseous layer of vaporized coolant. For the lower values of ΔT in the film boiling regime D, the heat flux drops back to values underneath 1 W/cm^2. The mechanisms of heat transfer for this region on the boiling curve are mainly composed by heat conduction and convective cooling inside the gaseous layer as well as the radiative heat transfer. The radiative heat transfer for this regime increases in significance towards higher values for ΔT [Kav01]. This is indicated by the non-linear increase on the heat flux in the double-logarithmic plot shown in Figure 3.1.

Alterations to the Heat Transfer of a Solid-Liquid Interface

For the design of a superconducting current limiting device with recovery-under-load characteristics, it is crucial to take into account the precise function for the maximum possible heat transfer from the superconductor to the surrounding liquid coolant. As already mentioned, the function for the heat transfer, depicted in Figure 3.1, depends on several parameters. Those being for example the properties of the coolant itself, the temperature of the coolant and the operation pressure of the system.
Preliminary investigations were carried out in order to determine the possibilities for improving the properties of the heat transfer from a super-heated superconductor surfaces into a surrounding bath of liquid nitrogen at ambient pressure. These investigations and the results are presented in the following.
For examining the heat flux from a solid body to the surrounding liquid nitrogen as a coolant, cubical-shaped copper samples with the dimensions of 30 x 30 x 30 mm^3 were used. The samples were equipped with a platinum temperature sensor (Pt100) for measuring the temperature in the center of the sample and determining the temperature difference between the sample surface and the adjacent liquid nitrogen [HN14]. Due to the high thermal mass of the copper samples and the good thermal conductivity k of copper, it was assumed that the surface temperature can be determined by measuring the core temperature while assuming a negligible temperature gradient inside the sample. This is possible for measurements on samples with a low surface area to heat capacity ratio.
To determine the heat transfer characteristics from the samples into the liquid nitrogen and to calculate the corresponding boiling curves, the samples were immersed into a open reservoir of LN_2 where they subsequently cooled down from room-temperature to the temperature of the liquid nitrogen bath (77.3 K). While cooling down, the sample temperature was measured and recorded.
From the recorded temperature versus time correspondence it is possible to calculate the heat transfer from the solid into the surrounding nitrogen bath using the heat equation

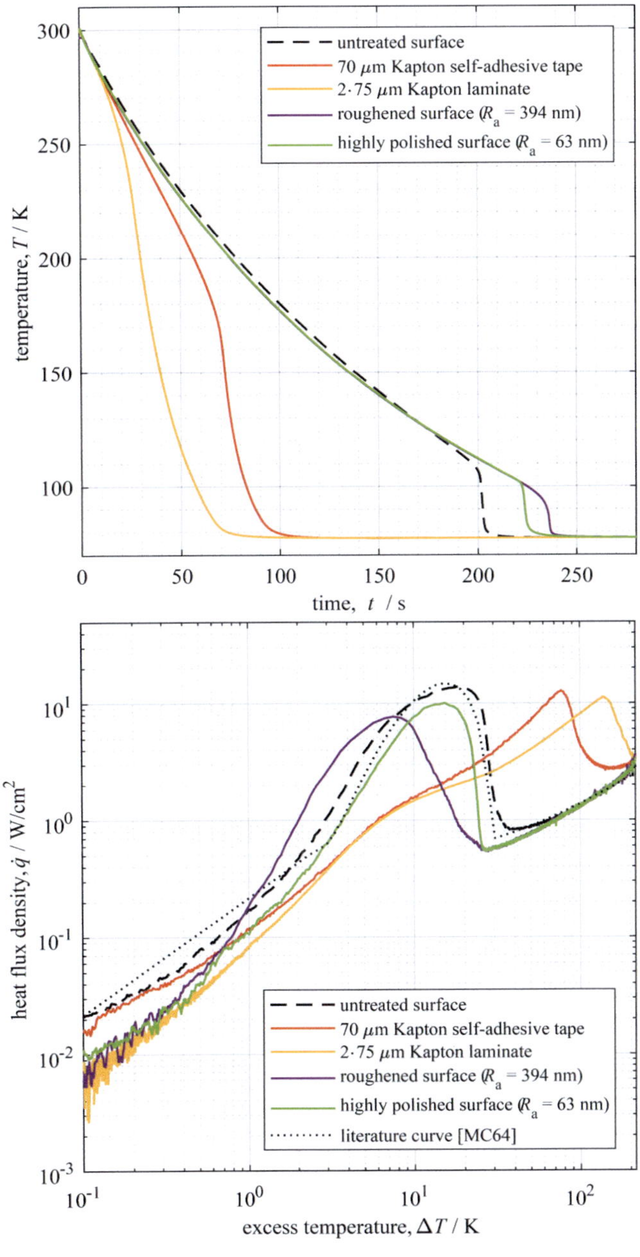

Figure 3.2: Influence of different surface treatments on the cool-down time and the boiling curve of 30 x 30 x 30 mm^3 copper samples immersed in LN_2 at 77.3 K and ambient pressure. The top graph shows the sample temperature versus the experiment time, the bottom graph presents the resulting alterations of the heat flux density versus the excess temperature (the presented results were published by S. Hellmann et. al. in [HN14])

stated in Equation 3.1.

$$\dot{q} = \rho \, c_{\mathrm{p}}(T) \, \frac{\mathrm{d}T}{\mathrm{d}t} \, \frac{V}{A_{\mathrm{surf.}}} \tag{3.1}$$

In Equation 3.1 ρ represents the density of copper and c_{p} the temperature dependent specific heat of the material, T is the measured temperature during the experiment as well as V and $A_{\mathrm{surf.}}$ represent the surface area and the sample volume respectively.
To investigate the cooling properties several samples were prepared. The samples had different surface adjustments such as finely polished surfaces of different qualities, artificially roughened surface in several grades of roughness and laminated surfaces with one or more layers of polyimide film (Kapton®). One sample without alterations to the surface was measured to provide reference values.
The measurement results are shown in Figure 3.2. In the top graph of this figure the sample temperature is shown versus the experiment time. The samples include the reference sample, two samples with Kapton® lamination on the sample surface, 70 µm of self adhesive Kapton® tape and 2 times 75 µm Kapton® laminate. Further depicted is a sample with a surface rougher than the surface of the reference sample as well as one sample with a highly polished surface. The average surface roughness R_{a} of those two samples was determined prior to the experiments using a profilometer. For the roughened sample R_{a} was measured as 394 nm and 63 nm for the polished sample.
From the graph it can be seen that the reference sample takes around 210 seconds to completely cool down from 300 K to 77.3 K (temperature of liquid nitrogen at ambient pressure). In comparison, the samples with the altered surface roughness need more time to cool to the temperature of the coolant, roughly 225 seconds for the polished sample as well as about 235 seconds for the sample with the artificially roughened surface. From the cool-down time alone it can already be determined that altering the surface roughness does not prove to have a positive effect on the efficiency of the heat transfer between the sample surface and the coolant. The two samples with laminated surfaces in the contrary show significantly reduced cool town times when immersed in the liquid nitrogen bath. Here the sample with one layer of 70 µm self adhesive Kapton® foil is cooled down to 77.3 K in about 100 seconds after the start of immersion into LN_2. The second sample with two layers of Kapton® laminate added to the surface reaches the coolant temperature level only about 80 seconds after the start of the experiment. This indicates a significant improvement to the heat exchange rate between the sample surface and the surrounding coolant [HN14].
The bottom graph of Figure 3.2 presents the boiling curves that were calculated from the recorded temperature versus time measurements on the samples immersed in liquid ni-

trogen with use of Equation 3.1. As a reference for the heat transfer the literature boiling curve determined by H. Merte et. al. [MC64] is depicted in this graph together with the measurement results. It shows that the sample with the untreated surface, measured for the work presented here, agrees very well with the literature data.

The samples with modified surfaces however show deviations to the literature boiling curve. The most significant deviation is hereby shown for the samples with additional Kapton® layers added to the sample surface. For these samples the boiling curves are shifted towards higher values for ΔT while the actual value for the heat flux remains in the same range. The sample with an additional layer of 70 µm self adhesive Kapton® film $\dot{q}_{max}$ experiences a shift to values between 70 K and 80 K excess temperature from an initial value of around 14 K. Hereby the range of efficient cooling around $\dot{q}_{max}$ is not only significantly shifted towards higher excess temperature ranges but also widened. This means, in turn, that the cooling of the sample operates in a more efficient cooling regime over a much wider span of excess temperatures. As a result of this, the overall cooling process from 300 K down to the coolant temperature of 77.3 K is faster, as it can be seen in the upper graph of 3.2.

For the sample with an additional laminated Kapton® layer with a thickness of $2 \cdot 75$ µm the resulting boiling curve is shifted even further and for this sample $\dot{q}_{max}$ resides at values for ΔT well above 100 K. The reason for this significant improvements of the heat exchange between the super-heated surface and the liquid coolant is the low thermal conductivity of the lamination material. This results in a steep temperature gradient inside the lamination material and therefore in a lower temperature difference ΔT between the outside of the laminate and the directly adjacent coolant [HN14]. Meaning that the additional low thermal conducting layers do not actually change the boiling curve of the interface between the solid, in this case the laminate, and the coolant, but they rather change the heat exchange for the whole system, from the copper surface to the laminate and into the coolant.

As it can already be expected from examining the results of the cool-down experiments presented in the top graph of 3.2, the boiling curves of the samples with alterations to the surface roughness do not experience widening or increase in the heat exchange (longer cool-down times). The calculated boiling curve for the sample with increased surface roughness (R_a = 394 nm) is shifted towards lower values for ΔT and more narrow in comparison with the reference and the literature sample. This makes the overall heat exchange less effective, what results in a longer cool-down time.

For the sample with polished sample surface (R_a = 63 nm) the shape and the positioning of the boiling curve remain close to the literature values and the reference sample but

the value of the heat flux generally decreases. The maximum heat flux measured for this sample is lower than 10 W/cm^2 in the point of $\dot{q}_{max}$.
It can be concluded that it is possible to alter the heat flux from solids into liquid coolants and to adjust it towards desired values, required for certain superconducting power devices and current limiters. The most effective alteration tested for this work, proved to be the addition of a thin, low thermal conductive layer to the surface of the solid. This can increase the heat exchange with the liquid coolant significantly and in turn decreases the recovery time of a superconducting device or even enable the chances for recovery where there was no recovery possible before.
However it should be mentioned, that the experiments presented in this chapter are quasi-stationary experiments and the acquired results are not directly transferable to transient applications. Their usability for fast transient heat exchange phenomenon depends on many parameters and needs to be examined for every application separately.

3.2 Superconductor Quench Characteristics

For the selection of a commercially available superconducting tape which meets the criteria defined by the transformer design described in Chapter 2, a set of preliminary measurements were conducted on the quench characteristics of different sample materials. These measurements include the fundamental determination of the critical current I_c of the superconducting tape to estimate the tapes current carrying capabilities, as well as basic measurements on the electrical resistance of the examined tapes. The resistance is measured versus the temperature. This temperature dependent resistance was used for updating the transformer model in order to predict the current limitation behavior of the tape after a quench and during temperature increase inside the material. With the measurement of the critical current and its fluctuation over different production batches of superconducting tapes, it can be determined how many parallel superconducting tapes are needed for a given transformer design to carry the nominal current and to guarantee a fail-safe operation during current limitation.
Figure 3.3 depicts typical I_c-measurements for two commercially available superconducting tapes. Both tapes are 4 mm wide *Re*BCO-thin-film tapes with different properties, herein after referred to as tape 1 and tape 2. The measurement shows a critical current of 151.7 A for tape 1 and 234.8 A for tape 2 respectively. This means that for the same transformer design a lower number of parallel tapes are needed for a superconducting winding equipped with tape 2 in comparison to a winding configuration utilizing tape 1, with a lower critical current density.

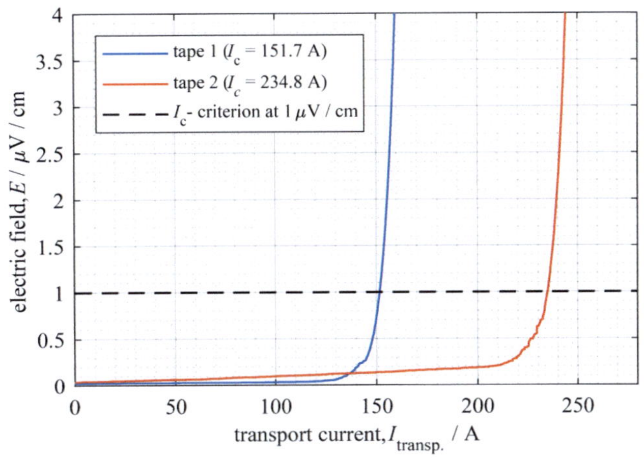

Figure 3.3: I_c-measurement for tape 1 and tape 2 . The measurement on tape 2 shows small indications for a current-transfer voltage, indicating an insufficient distance between the voltage taps and the current taps on the superconducting tape sample [Eki06]

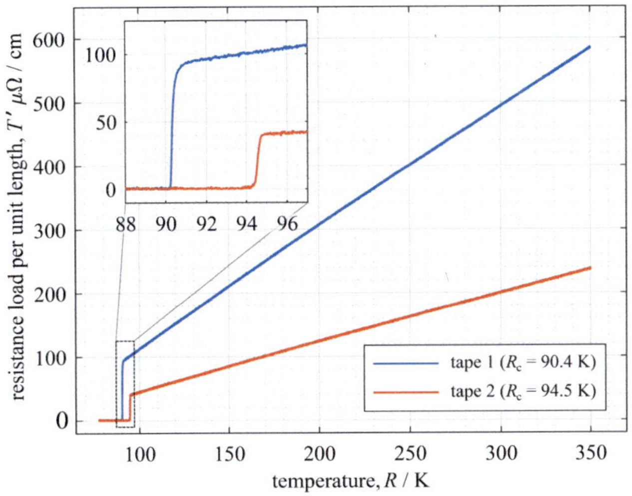

Figure 3.4: $R(T)$-measurement for tape 1 as well as for a tape 2 sample. For both samples the resistance was measured between 77.3 K and 350 K

The $R(T)$-measurement is mainly used for a current limiter design to determine the current limitation characteristics of superconducting tapes after quenching and during the warm-up of the material composition above the critical temperature T_c. Figure 3.4 shows a $R(T)$-measurement for tape 1 and tape 2. Both measured tape samples differ in the critical temperature by about 4 K, while tape 2 has the higher critical temperature with 94.5 K. Both tape samples show an almost linear temperature dependency of the resistance in the range between the critical temperature and the maximum measured temperature at 350 K. This is characteristic for the metallic normal conducting components in the tape composition (e.g. copper stabilization layer, protective silver layer and stainless steel substrate material). The proportional difference in the measured values for the normal conducting regime of both samples is caused by the different amount of normal conducting material in the layered tape structure. Tape 2 has a copper stabilization layer thickness of around 240 μm compared to a copper thickness of only around 110 μm for the material composition of tape 1. The higher amount of copper in the tape leads therefore to a lower resistance for tape 2 in the normal conducting regime.
The difference in the resistance of both tapes determines the efficiency of the current limitation of the superconducting transformer and potentially requires adjustment of transformer parameters in order to obtain the desired fault current limitation.
The transition of a superconducting material from its superconducting state into the normal conducting regime is a complex process and it depends on several parameters. These can include:

- the properties of the superconducting tape itself (I_c, T_c, the tape composition, the thickness and material of stabilization layers)
- the cause for the transition into the normal conducting state (current above I_c, temperature increase above T_c) and
- the amount of energy causing the transition from the superconducting to the normal conducting state.

The investigations on the heat transfer and the critical current as well as on the critical temperature of the superconducting tapes are not sufficient to reliably predict or calculate the transient quench process and therefore the transition into the current limitation regime. For this reason further experiments on the transition from superconducting state to the normal conducting state were carried out to determine the usability of the examined tape sample for the developed transformer design.
Those experiments were designed to simulate a fault current event as it would be experienced by the superconducting tapes if utilized in a current limiting transformer. During

the experiment the superconducting tape sample was immersed in a open bath of liquid nitrogen at ambient pressure and a predefined current profile was applied to the superconductor to emulate a nominal transport current as well as a short-circuit current. A typical current profile is shown in the top graph of Figure 3.5. Due to the use of an arbitrary signal power supply with a DC-output, the current profile was a rectified DC-current consisting of a nominal phase (with currents not exceeding I_c) and a limitation period (with currents above the critical current). While the predefined current profile was passed through the conductor the voltage drop along the sample was measured and recorded. The voltage drop corresponding to the depicted current profile is shown in the form of an electric field E in the bottom graph of Figure 3.5. The electric field along the conductor only shows very small values for the nominal current phase ($I_{\mathrm{nom,peak}} = 197$ A) before $t = 0$ s. In this regime the tape is fully superconducting and should not show any voltage drop, the small measured signal is rather caused due to induced voltages in the voltage measurement loop.

After the onset of the limitation period the current increases to peak values of 800 A. During this limitation period the superconductors critical current, indicated by a dashed line in the top graph of Figure 3.5, is exceeded and the superconductor experiences increased losses and therefore a resistive heat intake. This increases the superconductors temperature what in turn leads to a decrease in the critical current and eventually to a total breakdown of the superconductive state. The rising voltage drop along the superconducting tape is shown in the lower graph of Figure 3.5.

With the defined current as well as the measured voltage drop along the sample it is possible to calculate the resistance of the tape during the experiment. However, this resistance does not necessarily reflect the real resistivity value of the tape. This is because the tape can become partially normal conducting and therefore, the superconductor can show a significant increase of its local resistance while most parts of the sample are still at a temperature well below the critical temperature T_c.

The graph positioned at the top of Figure 3.6 shows the resistance load per unit length R' of the superconducting tape during the quench experiment, as it was calculated from the values presented in Figure 3.5. To avoid incorrect values in the calculated resistance, that are caused by dividing the measured voltage drop by sample current values close to zero, only sample current values above 200 A were taken into account for this analysis. The resistance load per unit length is shown as a black line in the top graph of Figure 3.6.

The measurement points, indicated by the circular markers in the same graph, also represent the mean resistance load per unit length for the tape. They are calculated by the peak values of the current divided by the corresponding peak values for the measured

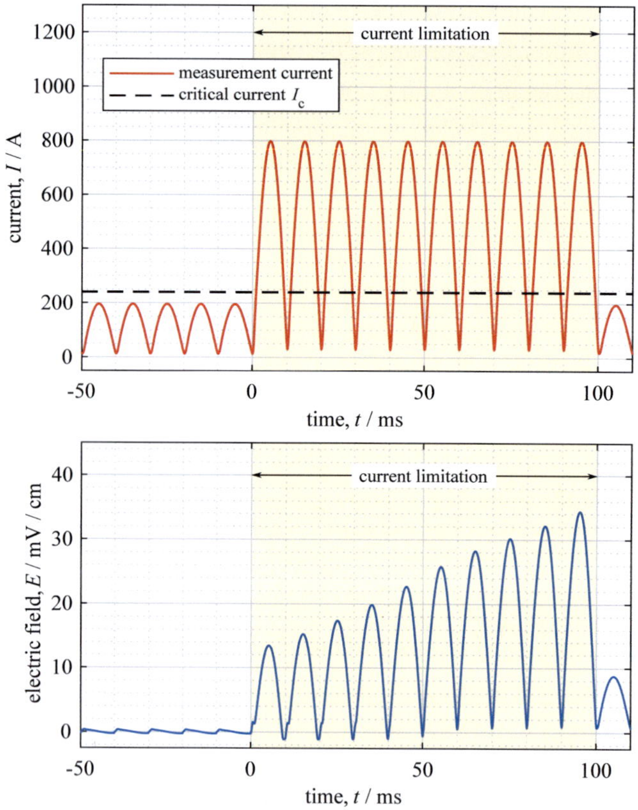

Figure 3.5: Given current profile (top) and measured consequent voltage drop along the sample (bottom) during a quench measurement on a conductor sample of tape 2

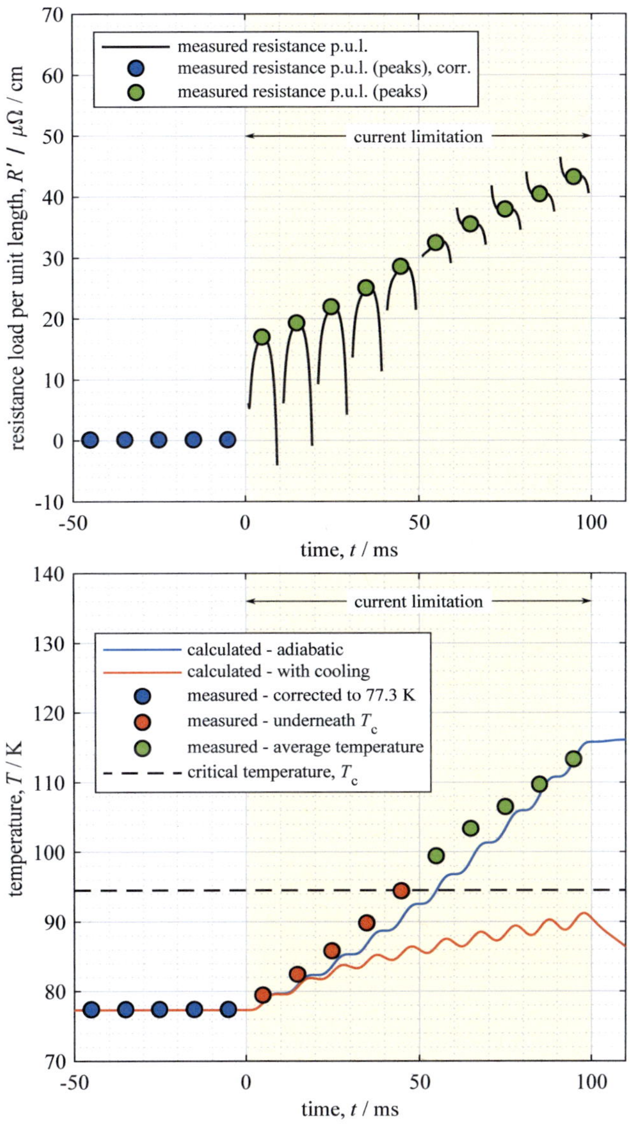

Figure 3.6: Transition of a tape 2 sample from the superconducting regime to the normal conducting regime. The top graph shows the resistance load per unit length R' whereas the bottom graph shows the temperature of the sample during the transition. The transition is caused by an over current of 800 A (peak), depicted in the top graph of Figure 3.5

voltage. The use of the peak values eliminates the small error in the calculated resistance that is caused by a phase shift due to inductive parts in the measurement circuit. Here the green markers show unaltered resistance values whereas the blue markers were corrected to zero because there is no voltage drop possible in the superconducting state.

It further has to be mentioned that the measured resistance for the superconducting tape represents an integral value of resistances load per unit length along the tape segment. From this measurement alone it can neither be concluded that the tape experiences a homogeneous increase in temperature nor can it be concluded that every point on the tape leaves the superconducting state at the same time.

The temperature of the superconducting tape is determined by using the previously acquired $R(T)$-measurement data. The sample temperature calculated with this method is depicted by circular markers in the bottom graph of Figure 3.6. Hereby the blue markers were again corrected to 77.3 K to compensate for the inductive error in the voltage measurement. The orange markers are based on resistance load values that indicate an incomplete quench of the tape and thus, a heterogeneous temperature distribution along the tape. This results in values for the calculated temperature that are below the critical temperature of the tape sample ($T_c = 94.5$ K). This temperature values can be interpreted as the average temperature of the resistive part of the examined tape.

The determined tape temperatures, shown by the green markers, are located above the critical temperature of the superconducting material and are assumed to represent a state of the quench where the superconductor is in the normal conducting regime over its full length. If this is the case, this measurement still represents the average temperature of the measured sample length and does not give any information on the actual temperature distribution or any specific local temperature.

For a better understanding of the quench process, a numerical simulation of the temperature increase in the measured sample was carried out. The simulation is based on a zero-dimensional Runge-Kutta approach, including the temperature dependent specific heat c_p of the superconductor material composition, optional cooling based on literature values [MC64] and the heat intake, calculated with the current and the voltage drop during the measurement.

The results of the simulation are represented by the continuous lines in Figure 3.6. The blue line represent the adiabatic temperature increase in the sample and the red line shows the temperature during the quench, including the cooling via adjacent liquid nitrogen. It can be seen that the adiabatic simulation of the temperature in the sample agrees with the directly calculated temperatures. The maximum deviation, between the measured temperature and the simulated result, appears around 50 ms after the onset of the fault

current with a value of about 5 K in difference. The reason for this is probably due to the non-uniformity of the quench propagation along the tape. Since the simulation model is of zero-dimensional nature, it does not include quench propagation but rather presumes a homogeneous increase in temperature along the sample length to all times.
It can be seen here that the non-adiabatic simulation result, including the heat transfer into liquid nitrogen according to H. Merte et. al., starts to deviate from the measured result, for temperature values 30 ms after the onset of the short-circuit current. The non-adiabatic simulation suggests that the sample temperature stays below the critical tape temperature T_c for the whole fault duration. The reason for this is the fact that the boiling curves measured by H. Merte et. al. are quasi-stationary results and therefore cannot be applied to fast transient quenching events. These fast transitions of superconducting tapes into the normal conducting regime should rather be considered adiabatic. While this does not apply for all superconductor quench processes, it was found to fit well for the quench behavior of the superconducting tapes, as discussed in this work.

3.3 Optical Quench and Recovery Investigation

During the measurements on the heat transfer from solids to liquid coolants (Chapter 3.1) and the investigation on the fast transitions to the normal conducting regime (Chapter 3.2) it became clear that the quench and recovery process of a superconductor requires further investigations towards a full understanding. Thus it was decided to not only perform electrical and/or thermal measurements on superconducting tape samples but also to use an optical measurement method in order better understand the experimental results and to directly observe boiling phenomena, or to distinguish different boiling phases on super-heated surfaces adjacent to liquid nitrogen.
This was implemented with a high-speed camera setup capable of providing high frame rate video sequences of quenching and recovering superconductor samples and general heat transfer processes inside a open cryostat with liquid nitrogen as a coolant. A more detailed explanation on the experimental setup itself is given in Chapter 8.1.
A set of video frames taken of a quenching superconductor sample is shown in Figure 3.7. The plotted frames are extracted from the video sequence and present a side view of the full superconducting sample length at different times during the experiment. On the left and on the right hand side of the depicted frames the current contacts show partially as well as the voltage taps soldered to the tape surface can be seen. The frames shown in Figure 3.7 present the superconducting tape sample in normal operation (frame A), during quench propagation and full quench (frame B and C respectively), during

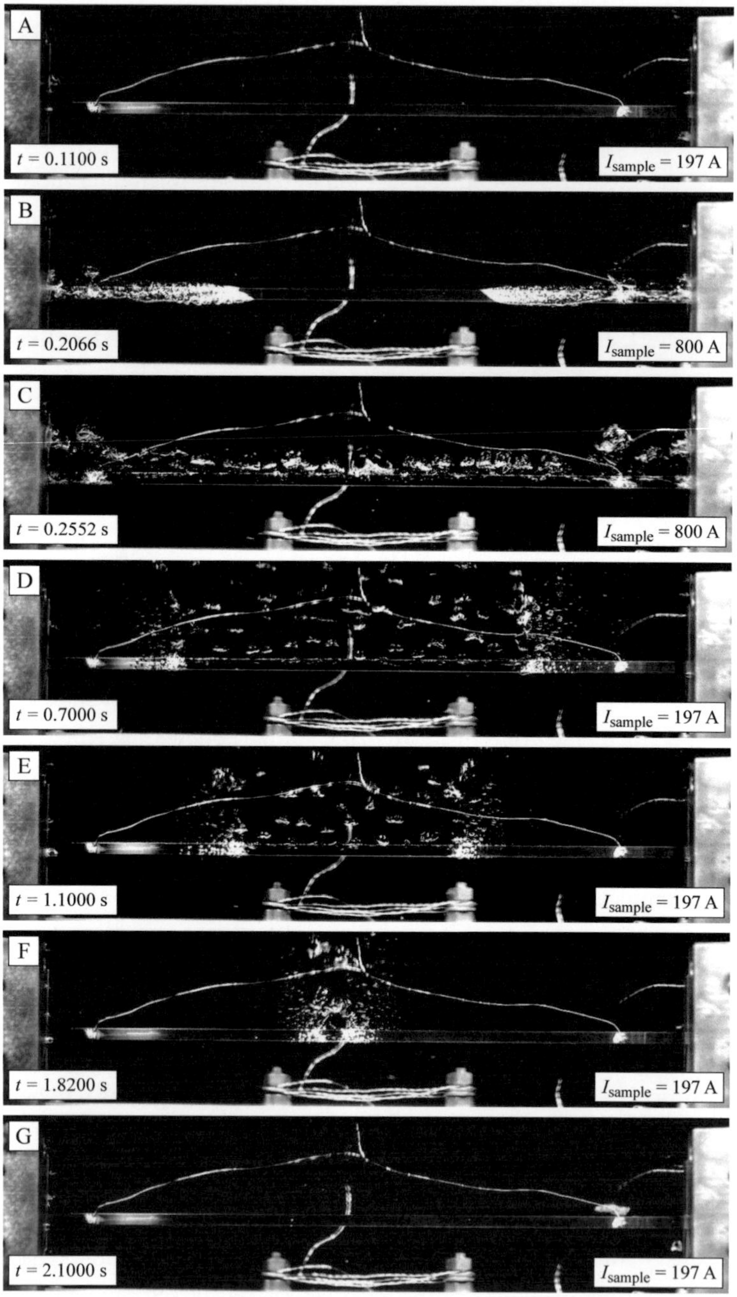

Figure 3.7: Different frames taken from a high-speed video recording of a quenching, 4 mm wide superconducting tape (tape 2). The frames are acquired at different times, as shown on the lower left side of each frame. Frame A, B and C show the tape before and during the quench, frame C, D, E and F are taken during the recovery-under-load process and frame G shows the fully recovered tape (the presented results were published by S. Hellmann et. al. at the European Conference for Applied Superconductivity 2015)

recovery-under-load (frame D, E and F) and fully recovered back to nominal operation (frame G). The fault current in this experiment lasted for 100 ms with peak values of 800 A whereas the nominal current and the recovery current were set to 197 A (peak).
The high-speed video recordings proved to be a valuable tool in the investigation on the quench and recovery processes of superconducting tapes. Especially if an evaluation is carried out in combination with electrical measurements that are acquired synchronized to the recorded video data.
For an better comparison of the video data, the frames taken during an experiment were composed into a single array as a part of the data post processing. The methodology behind the composition of this single array, representing the whole video data, is depicted in Figure 3.8.
Section A of Figure 3.8 shows a raw video frame with the the different boiling phases along the tape sample marked in green color. For the data composition only a single line of pixels, marked as a red dashed line, is extracted from each original video frame and composed into the new array as shown in section B of Figure 3.8 and as indicated by the red arrow. The extracted pixel lines are arranged in chronological order and the resulting array allows to distinguish the different boiling phases on the measured tape sample versus the experiment time.
The film boiling phase is shown as a darker area in the composed array because the gaseous layer of evaporated nitrogen on the tape deviates the incoming light. This darkens the appearance of the tape and therefore the intensity in the extracted row of pixels. The nucleate boiling and transition boiling produce a higher number of bubbles directly on the tape surface. Those bubbles cause reflections of the incoming light into the camera lens and they consequently appear as brighter areas in the composed array. However it is not easily possible to distinguish between areas of nucleate boiling regime and transition boiling regime in the composed array. The recovered areas on the tape as well as the areas with no boiling, appear unaltered and directly represent the tape surface. The evaluation of the bubble coverage and the development of the different boiling phases allows for conclusions on the local temperature of the tape sample as well as on general quench and recovery characteristics and quench propagation along the tape.
Also visible in the resulting array are the solder dots connecting the voltage taps to the tape surface, they are marked in orange color in Figure 3.8. Since the array composed according to this method is visualizing the bubble coverage during the different boiling phases, herein after this is referred to as *bubble coverage array*.

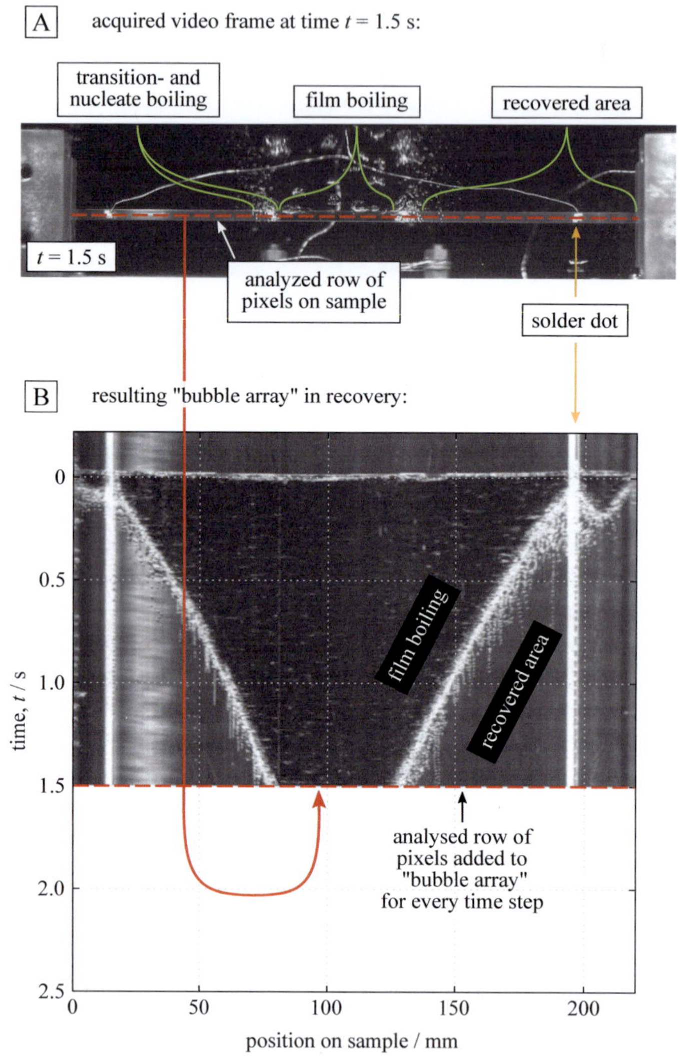

Figure 3.8: For a simple comparison of the recorded videos, only single lines of pixels on the tape surface were analyzed in each video frame. Those lines were rearranged into arrays, visualizing the bubble coverage of the superconducting tapes during a quench and recovery experiment. This gives information of the local tape temperature as well as on the different boiling phases during the experiment

3.4 Recovery-under-Load Characteristics

For a superconducting current limiting device, especially the recovery-under-load behavior is of interest, since a successful recovery-under-load can allow for a efficient current limitation with eliminating the need for the current limiter to be disconnected after clearing the fault from the power grid. The load of the device during the recovery phase can be full or partial nominal load.

The bubble coverage array shown in Figure 3.9 represents the high-speed video recording acquired during the quench measurement that was already discussed in Chapter 3.2, with Figure 3.5 and Figure 3.6. Whereas in Chapter 3.2 the focus in this measurement was on the quench behavior, the focus for the results presented here, is on the recovery-under-load characteristics. From Figure 3.9 it can be seen that at the end of the quench (t = 0.1 s) the superconducting tape is in the film boiling regime over its full length. This is indicated by the darker areas of the bubble coverage array. It can be seen that the recovery-under-load process starts from the current contacts and the solder dots on the tape, and evolves towards the middle of the sample. During the recovery phase, the last film boiling area disappears about 1.7 s after the onset of the fault current, close to 100 mm from the left side of the tape sample. With the disappearance of the film boiling area, it can be concluded that the maximum temperature of the tape is now lower than 28 K above the liquid nitrogen temperature of 77.3 K. This is possible because the onset of the film boiling regime is given with a ΔT of 28 K in the boiling curve by H. Merte et. al.

Several optical measurements on the recovery of superconducting tapes showed that the recovery process always starts from initial points on the tape surface with increased specific heat capacity. Those were for example the copper current contacts as well as the solder dots for contacting the voltage measurement taps. The reason for this is the additional heat capacity at those points, what keeps the temperature during the quench of the tape sample at lower values compared to the bare parts of the superconducting tape. In an ideal case the excess temperature after the quench is kept underneath the temperature at $\dot{q}_{\mathrm{max}}$ in the boiling curve. For this reason recovery starts with an significantly more efficient heat transfer in those areas compared to the areas that passed the excess temperature of $\dot{q}_{\mathrm{max}}$ and fully went into the film boiling regime.

The array in Figure 3.9 shows only two points where the tape recovers back from film boiling, via transition boiling and nucleate boiling. Those points are observed to the left and to the right hand side of the dark film boiling areas in Figure 3.9 and form a triangular shaped film boiling pattern during the recovery process of the tape.

After this observation identical tape samples were modified with additional solder dots

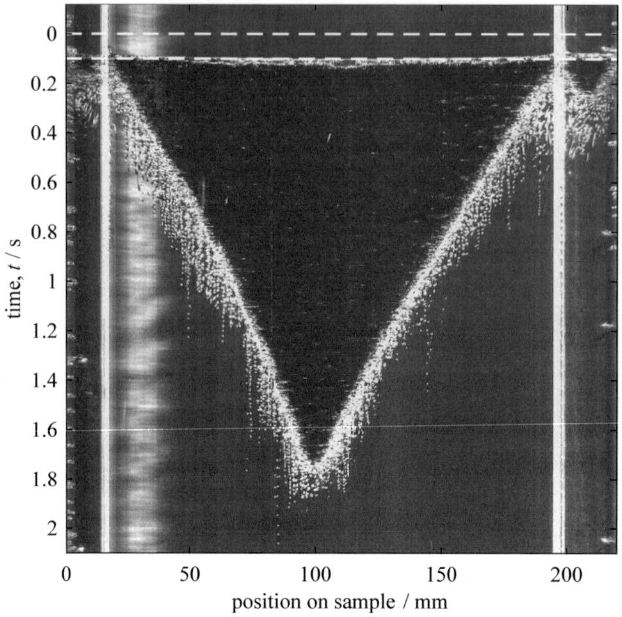

Figure 3.9: Bubble coverage array of a superconducting tape sample (tape 2) during recovering under load ($I_{\text{rec.,peak}}$ = 197 A). The fault current causing the quench was $I_{\text{lim.,peak}}$ = 800 A for a duration of 100 ms. The fault duration is indicated by the two dashed lines in the depicted array. The critical current of the tape is 225 A (the presented results were published by S. Hellmann et. al. at the European Conference for Applied Superconductivity 2015)

in order to provide local points with increased specific heat along the tape and to act as initial recovery points. Figure 3.10 shows composed bubble coverage arrays of measurements on tapes with four and six solder dots applied to the tape surface. While the peak limitation current ($I_{\text{lim.,peak}}$ = 800 A for 100 ms) and the peak recovery current ($I_{\text{rec.,peak}}$ = 197 A) during the additional tests remained the same, the recovery times for the samples with four and six solder dots decreased significantly. For the sample with four solder dots a full recovery-under-load can be observed 0.7 s after the onset of the limitation current. For the tape sample with six solder dots applied to the tape surface, the recovery time decreases further and reaches a full recovery-under-load less than 0.25 s after the onset of the limitation current.

The reason for this significant decrease in the recovery time are the additional points for an initial onset of recovery, provided by the locally increased heat capacity of the solder dots. Instead of having only two points on the tape sample for the transition from film boiling back to a recovered state, every additional solder adds two more transition points along the tape, as it can be observed in both graphs of Figure 3.10. This significantly increases the efficiency of the heat transfer from the tape surface to the surrounding

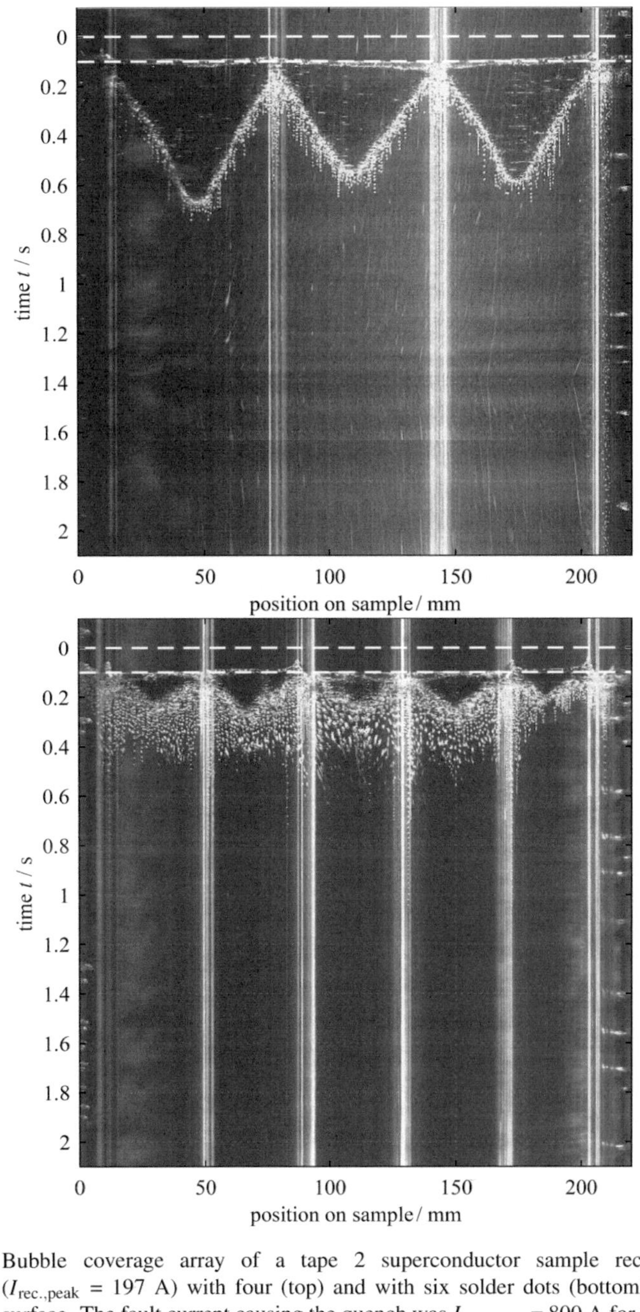

Figure 3.10: Bubble coverage array of a tape 2 superconductor sample recovering under load ($I_{\mathrm{rec.,peak}}$ = 197 A) with four (top) and with six solder dots (bottom) applied to the tape surface. The fault current causing the quench was $I_{\mathrm{lim.,peak}}$ = 800 A for a duration of 100 ms. The fault duration is indicated by the two dashed lines in the array. The critical current of both tapes is 225 A (the presented results were published by S. Hellmann et. al. at the European Conference for Applied Superconductivity 2015)

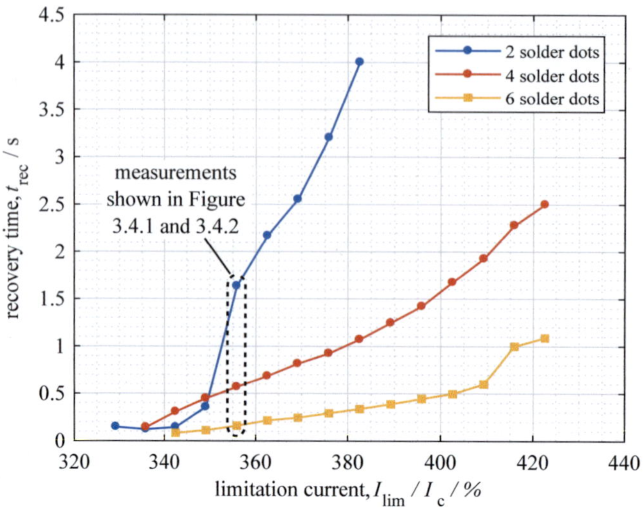

Figure 3.11: Reduction in recovery-under-load time achieved by adding solder dots to the tape surface (tape 2). The limitation current is specified as a percentage of the critical current ($I_c = 225\,A$) (the presented results were published by S. Hellmann et. al. at the European Conference for Applied Superconductivity 2015)

liquid nitrogen and therefore decreases the recovery time.

Figure 3.11 shows the effect of the additional solder dots on the recovery time for the three samples presented in Figure 3.9 and Figure 3.10. The graph visualizes the recovery-under-load time versus the limitation current. The limitation current is given as a percentage of the critical current. It can be seen that, above a certain level for the limitation current, the addition of further solder dots continuously shortens the recovery times. The blue line represents the measurements conducted for the sample with two solder dots and it ends just above a limitation current value of 380 % of I_c. For higher values of the limitation current the sample experienced a thermal runaway and a consequent thermal degradation. Whereas the samples with four and six solder dots can fully recover for limitation currents above 380 % of the critical current. The yellow line, for the sample with six solder dots, shows a step to longer recovery times for the last two recorded measurements. For those two measurements the heat capacity of the additional solder mass proved to be too low and several solder dots heated up above the excess temperature of $\dot{q}_{max}$ and therefore reached the film boiling regime. Thus they were not acting as seeds for recovering areas anymore but behave like bare parts of the tape. This effect shows, that the value of the additional thermal mass has to be adjusted for every application in order to improve the recovery-under-load behavior in a reliable manner.

It should further be mentioned that the addition of solder dots can, depending on the con-

ductivity of the used solder and the dimension of the applied solder dots, decrease the normal conducting resistance of the tape composition and therefore also influence the current limiting behavior. But, since the improvement in the recovery-under-load behavior by additional local heat capacities is not depending on the electrical conductivity of the added materials, these material can also be made of a non-conducting material.
In the examples presented here, the solder dot mass is between 70 mg and 90 mg. This decreased the current limitation behavior of the configuration with six solder dots, compared to the configuration with two solder dots, by less than 2 %.
Additionally to the measurements, a numerical simulation of the recovery-under-load characteristics is applied to reproduce and examine the recovery process. The simulation model and the simulation results are presented in Chapter 8.2.
The additionally added solder dots on the surface of the superconducting tapes show promising reductions in the recovery times after a quench of the superconductor. However, this method for reducing recovery times is not applied for the design of the current limiting transformer demonstrator described during the course of this thesis.

3.5 Summary of Preliminary Experimental Investigations

For all commercially available tapes, a recovery limit was determined in order to find the best suited tape for the application in the superconducting fault current limiting transformer, as introduced in Chapter 2. The recovery limit was defined as the maximum temperature a conductor is allowed to reach during the current limitation phase, while still being capable of recovering under the transformers full nominal load.
Figure 3.12 shows the recovery limits for the examined conductors. The measurements for composing the curves were conducted by gradually increasing the conductor temperature, reached after the current limitation (T_{max}), by increasing the value of the fault current. The current limitation phase was consistent with the previously conducted measurements and composed of ten half-waves of sinusoidal current for a duration of 100 ms. The measurements shown for a single conductor in Figure 3.12 give the maximum temperature after the current limitation (y-axis), followed by a recovery-under-load phase with a certain recovery current (x-axis) that allowed successful recovery. For higher maximum temperatures the conductor experiences a thermal runaway and irreversible thermal degradation during the recovery phase. Here the recovery current is given as a percentage of the critical current I_{c} to enable a comparison between conductors with different critical currents.

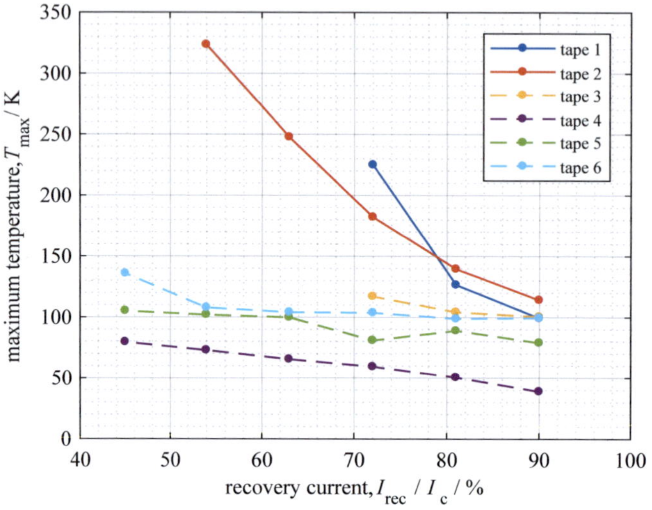

Figure 3.12: Recovery limits for different commercially available superconducting tapes. Maximum temperature above the indicated lines are to high for allowing a successful recovery-under-load after a quench. The depicted lines can therefore be referred to as recovery limits

From the graph it can be seen that two of the determined curves, describing the recovery limit, reach into higher temperature regions for T_{max}. These two conductors (tape 1 and tape 2) are of higher interest for the suggested transformer design, because the transient simulation described in Chapter 2 predicted maximum temperatures after the current limitations in the range between 200 K and 300 K. In the suggested transformer design the recovery current for full nominal load of the transformer corresponds with values for $I_{rec.}$ / I_c of between 60 % and 70 %.

The reason for the higher tolerance in the maximum temperature of tape 1 and tape 2 is the increased amount of thermal and electrical stabilization used in the tape composition. This results in a lower resistance of the superconducting tape after the quench and therefore less resistive heating during the recovery phase.

In Figure 3.12 tape 4 and tape 5 show recovery limits with a maximum temperature values below their respective critical temperature (see Table 8.1 in Chapter 8.3). This indicates a inhomogeneous or incomplete quench of the superconducting tape during the recovery-under-load measurements.

After the the preliminary measurements and the determination of the limits for recovery-under-load, tape 1 showed the most promising characteristics for the application in the superconducting transformer. The best alternative for this tape would have been provided by tape 2. The detailed specifics for both tapes are given in Table 3.1. A table with the full specifications of all six evaluated tapes can be found in Chapter 8.3.

Table 3.1: Specific properties for tape 1 (main option) and tape 2 (alternative option for the transformer design). The tapes were selected based on the preliminary measurements . Given are the values as stated by the manufacturer whereas the values in parenthesis are measurement results

property	tape 1	tape 2
manufacturer	SuperPower® Inc.	SuNAM® Co. Ltd.
critical current	> 140 A (151.4 A)	> 200 A (234.2 A)
critical current variation	± 10 %	± 10 %
critical temperature	- (90.4 K)	- (94.5 K)
copper stabilizer, electroplated	110 μm (105.3 μm)	30 μm (30 μm)
copper stabilizer, laminated	- no laminated copper -	200 μm (185.8 μm)
silver layer thickness	2 μm	2 μm
substrate thickness	50 μm (48.0 μm)	95 μm (95.5 μm)
*Re*BCO-layer thickness	1 μm	2 μm
total tape thickness	163 μm (160.8 μm)	350 μm (351.5 μm)
tape width	4 mm	4 mm
minimum bending radius	< 50 mm	< 50 mm

In addition to the performance test on all available tapes during the preliminary measurements, all commercially available tape structures (tape 1 to tape 6) were also assessed during part 2 of the transformer design phase as described in Chapter 2.2.2.

The simulated transient results confirmed, that tape 1 is the best candidates for the superconducting transformer that is designed and manufactured during the course of this work.

4 Components and Manufacturing of a 1 MVA-Class Current Limiting Transformer

This chapter gives a general understanding of the manufacturing and assembly process of the components in the superconducting transformer demonstrator. The transformer design is based on the specifications and design results as they are listed in Chapter 2.3. Chapter 4.1 gives an overview of the construction of the transformer. In contrary to the transformer features described in Chapter 2.1 this overview is focused on the engineering details and the manufacturing processes necessary for the assembly of the superconducting transformer demonstrator. The Chapters 4.2, 4.3, 4.4 and 4.5 give a more detailed explanation on the main transformer components. These are the laminated iron core, the normal conducting primary winding, the transformer cryostat and the superconducting secondary transformer winding.

4.1 Overall Transformer Setup

The transformer setup is based on the dimensions and specification resulting from the transformer design process as given in Chapter 2.3.

Figure 4.1 shows the main components of the transformer for the laboratory test setup prior to the assembly. Depicted on the left hand side of the photography is the bottom part of the iron core together with the primary winding. The transformer cryostat is already inserted into the primary winding. The bottom part of the cryostat assembly provides the vacuum port which is connected to the vacuum pumping system. At the top end of the cryostat, the ring-shaped LN_2-gap is visible.

The primary winding as well as the transformer cryostat are arranged axially symmetric around the core limb. Both parts are sitting on isolated spacers, resting on the lower part of the iron core clamping bars. Since the primary winding as well as the cryostat are not mechanically fixed, a good axial symmetry between the windings is required to minimize the magnetic forces during normal operation and especially during current limitation.

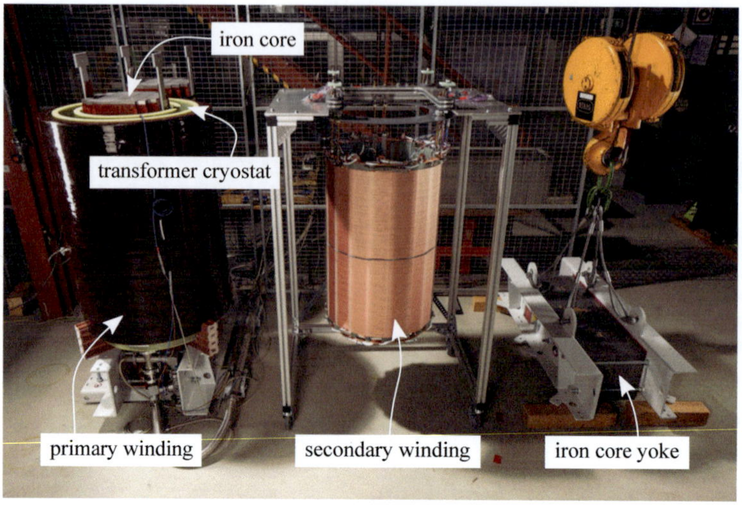

Figure 4.1: The photography gives an overview of the different transformer components. Here the superconducting secondary winding is sitting in a temporary rack during assembly

In the center of Figure 4.1 the fully assembled superconducting secondary winding is shown. The secondary winding is suspended from the aluminium busbars above and outside the transformer cryostat. The aluminium busbars in turn are mechanically fixed to the core clamping bars of the iron core yoke. The winding suspension is realized using the current leads, described in Chapter 4.5. The secondary winding, as shown in the photography, is fully equipped with all measurement accessories.
The iron core yoke shown on the right of the photography in Figure 4.1 is placed on top of both iron core limbs. The yoke closes the magnetic flux circuit in the core and is added as the last step of the transformer assembly.
To give a better understanding of the transformer geometry, Figure 4.2 shows a cutaway view of the fully assembled transformer. This computer aided visualization is based on the original computer aided design (CAD) drawings that were used during construction and for the machining of the transformer parts.

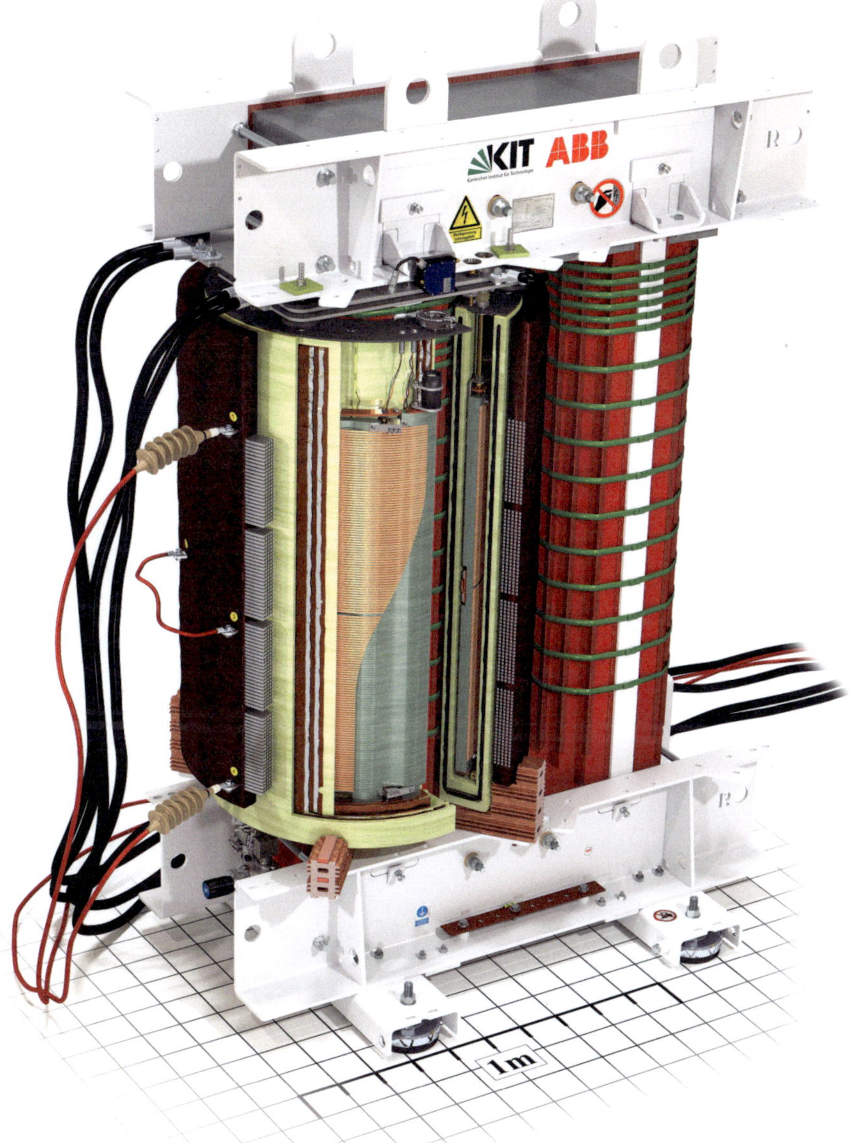

Figure 4.2: Cutaway view of the fully assembled superconducting current limiting transformer (computer aided visualization)

4.2 Iron Core

The iron core of the transformer demonstrator is of a U-I core type. It consists of two parts, a U-shaped bottom part composed by the two core limbs and the bottom core yoke, and a removable top yoke. For this transformer, only one of the limbs is the wound limb whereas the other limb does not carry any windings and purely acts as a magnetic flux return path.

Both the transformer core parts are made from laminated sheet metal. The single sheets have a thickness of 300 μm and are cut from grain oriented core steel manufactured by Thyssen Krupp - Electrical Steel GmbH and sold under the brand name powercore® C 140-30. The sheets are electrically insulated on both sides by a layer of glass with a thickness between 2 μm and 5 μm. The electrical insulation is necessary to avoid the formation of large eddy-current loops in the stacked iron ore sheets.

The core material has a linear magnetization regime up to a magnetic flux density of about 2 T [Thy18]. For the maximum flux density in the iron core B_{max}, specified in Chapter 2.3 as 1.6 T, the core losses are given by the manufacturer as 1.01 W / kg per kilogram of core material, assuming a sinusoidal magnetization with a frequency of 50 Hz.

The total core mass of the transformer is 4001.3 kg, which does not include the core clamping bars or any other structural core parts. With magnetic losses of 1.01 W / kg at 50 Hz this leads to total magnetization losses in the core of 4041.3 W. The specific core material losses as stated by the manufacturer of the iron core material are presented in the graph shown in Figure 4.3.

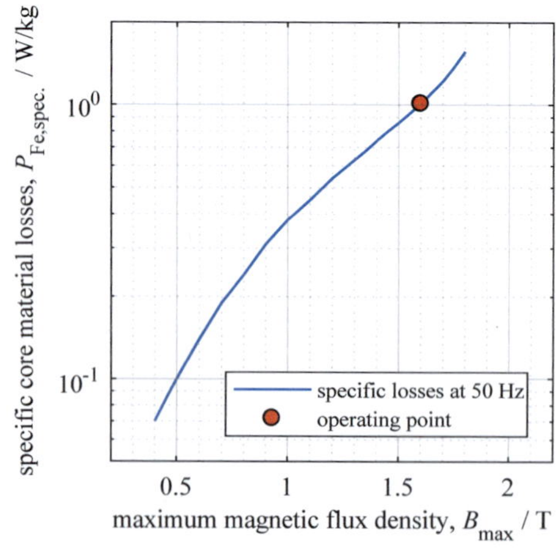

Figure 4.3: The graph shows the specific losses of the iron core material versus a sinusoidal core magnetization at 50 Hz. The values are determined by the manufacturer of the iron core material [Thy18]

Table 4.1: Iron core properties and iron core material properties. Values marked with * are extracted from [Thy18]

property	symbol	value
iron core mass (including core clamping frame)	m_{Fe}	4001.3 kg (~5100 kg)
iron core height (including core clamping frame)	-	1952 mm (2412 mm)
iron core depth (including core clamping frame)	-	366 mm (980 mm)
iron core width (including core clamping frame)	-	1022 mm (1500 mm)
total iron core cross section area	A_{Fe}	0.1185 m^2
effective iron core cross section area	$A_{\mathrm{Fe,eff.}}$	0.1144 m^2
core stacking factor	k_{Fe}	0.965 * (96.5 %)
iron core window height	-	1310 mm
iron core window width	-	280 mm
iron core material	-	powercore® C 140-30
iron core material density	ρ_{Fe}	7650 kg / m^3 *
specific iron core material losses for max. flux density of 1.6 T at 50 Hz	$P_{\mathrm{Fe,spec.}}$	1.01 W / kg *
material manufacturer	-	Thyssen Krupp - Electrical Steel GmbH
maximum flux density before saturation	-	2.03 T *
number of steps in iron core limbs	-	3
iron core manufacturer	-	ABB AG, Brilon, Germany

The properties of the iron core and the iron core material are listed in Table 4.1. A computer generated visualization of the iron core without any additional assembly is shown in Figure 4.4.

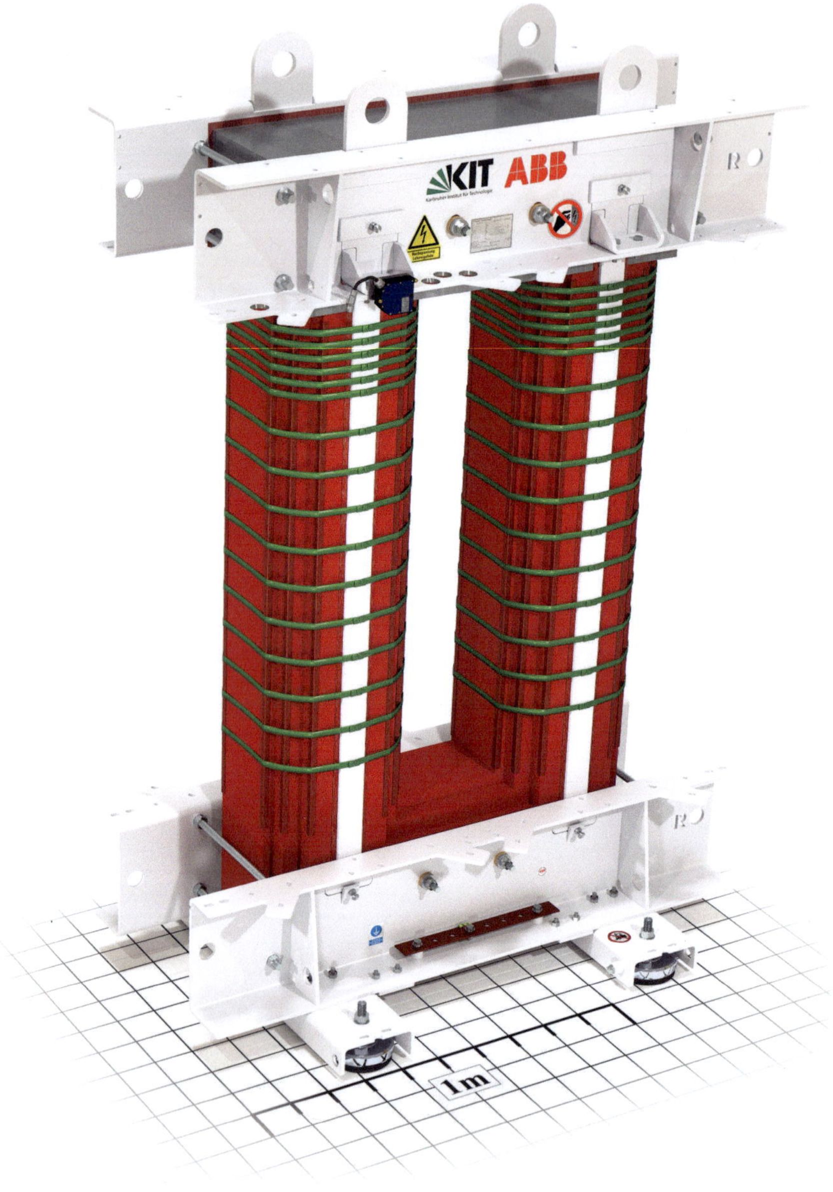

Figure 4.4: Conventional transformer core made from laminated grain oriented electrical steel. The core assembly has a removable core yoke to allow access to the transformer windings (computer aided visualization)

4.3 Primary Winding

The primary winding of the transformer is build as a resin-cast normal conducting winding as it is widely used in conventional transformers up to a power-class of about 40 MVA. The main advantage of this type of transformer winding is the commercial availability, the possibility for convective air cooling as well as the good insulation properties of the resin material. Especially the electrical insulation properties allow a compact assembly with only a small clearance between the inner side of the primary winding and the outer cryostat wall.

The total mechanical thickness of the winding is 46 mm, from the surface of the inner insulation to the surface of the insulation on the outside of the winding. The inner electrical diameter of the winding is 740 mm. The winding is wound of 500 turns of aluminium conductor with a cross section area of 31.5 mm^2. This results in a filling factor for the primary winding cross section area of 35.4 %. The resistive losses in the primary winding are expected to be around 885 W at the rated current of 28.9 A. To provide a good magnetic coupling and therefore a low short-circuit impedance value u_k, the primary winding has the same electrical height of 905 mm as the superconducting secondary winding.

A computer aided visualization of the winding is shown in Figure 4.5. It can be seen that the conductors in the primary winding are arranged in packs rather than equally distributed over the height of the winding. In resin-cast windings this is often done for insulation and manufacturing purposes or to add additional electrical taps along the winding.

It should be mentioned that an inhomogeneous arrangement of the conductors over the cross section area of the primary winding can influence the AC-losses caused in the secondary winding. To guarantee low AC-losses, the magnetic field in the stray gap, generated by the primary winding, should be as homogeneous as possible over the height of the winding.

The properties of the primary winding are listed in Table 4.2.

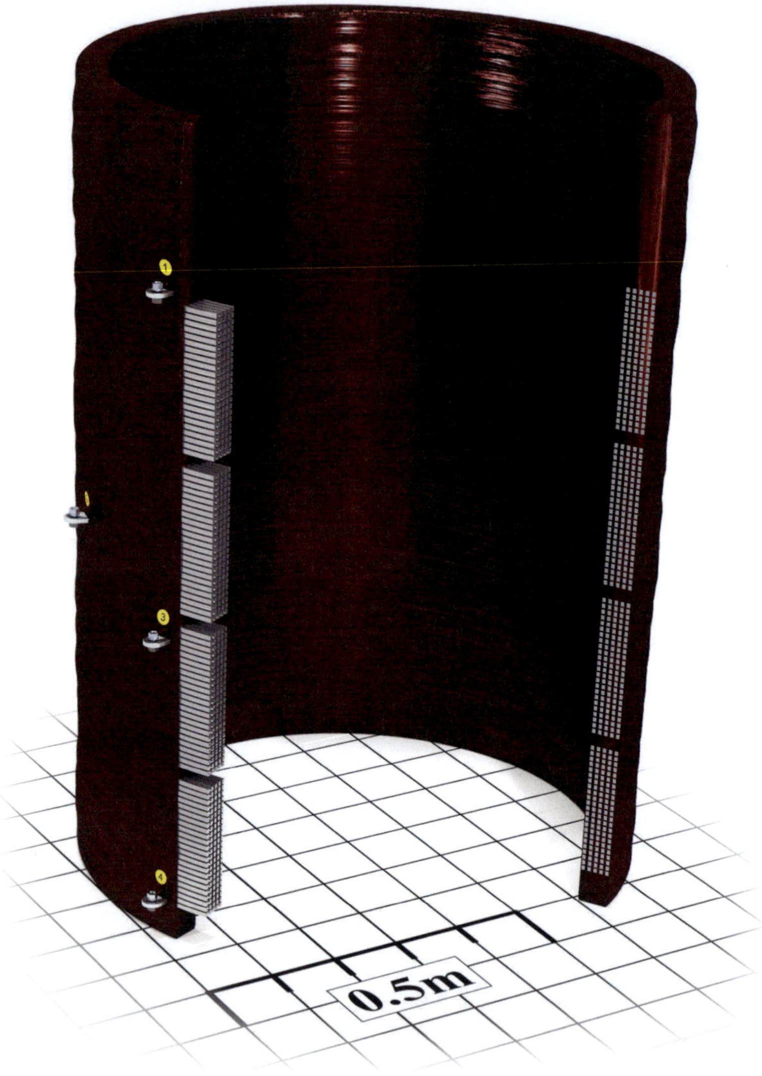

Figure 4.5: Sectional view of the conventional resin-cast 20 kV primary winding (computer aided visualization)

Table 4.2: Properties of the resin-cast primary transformer winding

property	symbol	value
nominal operation current	I_p	28.9 A
nominal operation voltage	U_p	20 kV
turn-to-turn voltage	u_w	40 V
primary winding weight (calculated)	-	296.4 kg
primary winding height (including insulation)	$h_{w,p}$	905 mm (1195 mm)
inner radius (including insulation)	-	370 mm (368.5 mm)
outer radius (including insulation)	-	412 mm (414 mm)
winding thickness (including insulation)	-	42 mm (45.5 mm)
conductor material	-	Aluminium
conductor height	$h_{c,p}$	6.3 mm
conductor width	$w_{c,p}$	5 mm
turns in winding	N_p	500
primary conductor length	-	1194.7 m
room-temperature resistance	R_p	1,126 Ω
winding inductance (calculated with [Whe28])	-	120.0 mH (error < 1 %)
filler insulation on top	-	230 mm
filler insulation at bottom	-	60 mm
primary winding manufacturer	-	ABB AG, Brilon, Germany

4.4 Transformer Cryostat

The cryostat encloses the superconducting current limiting winding of the transformer. In order to achieve good magnetic coupling between both windings, the stray gap of the transformer needs to be as narrow as possible. This can only be achieved with a vacuum insulated cryostat in combination with multi-layer insulation (MLI) inside the vacuum gap.

The multi-layer insulation consists of several stacked layers of aluminium coated Mylar or Kapton®-films. Each of these layers has a typical thickness of 10 µm or less and the layers are interlaced with thermally low conductive spacers to reduce heat conduction. The stacked layers provide a very effective method for reducing the radiative heat transfer through the vacuum gap of a cryostat [FSS03].

In the described cryostat design a total of two multi-layer insulation stacks, consisting of ten layers of coated foils each, are used in the inner and outer cryostat wall as well as in the bottom vacuum gap of the cryostat.

The cryostat cannot be made from electrically conductive material since this would form a conductive loop around the transformer main magnetic flux path and thus, result in a induced voltage and subsequent current flow in the cryostat material. For this reason a cryostat made of glass fibre reinforced plastic (GFRP) was designed. It offers good thermal insulation properties with a total inner and outer wall thicknesses of only 28 mm and 29 mm. Further, the material is non-conductive and therefore applicable for applications in a magnetic AC-field.

For the cryostat construction, the multi-layer insulation as well should not form a conductive loop around the transformer iron core. Therefore the multi-layer insulation in the cryostat walls is additionally interlaced with electrically insulating layers of Kapton®-foil to avoid conductive loops around the main magnetic flux path.

Figure 4.6 presents a cutaway view of the cryostat. The multi-layer insulation can be seen inside the vacuum gap in the inner and outer cryostat walls as well as in the bottom part. The cutaway also shows the vacuum flange at the bottom of the cryostat. This is connected to a vacuum pumping system and used for the evacuation of the vacuum gap and is required for increasing the insulation properties of the multi-layer insulation [FSS03]. Due to the high rate of outgassing, which is characteristic for GFRP-materials, the cryostat has to be frequently evacuated to provide the necessary quality of thermal insulation.

The cryostat is operated as an open cryostat with liquid nitrogen at ambient pressure. It can hold up to 125 l of LN_2 at a maximum filling level of 100 mm underneath the upper cryostat rim.

Table 4.3: List of properties of the vacuum insulated transformer cryostat

property	symbol	value
cryostat weight	-	170 kg
total cryostat height	-	1245 mm
height of LN_2-gap	-	1180 mm
inner radius	-	229 mm
outer radius	-	354 mm
LN_2-gap width	-	68 mm
thickness inner cryostat wall	-	28 mm
thickness outer cryostat wall	-	29 mm
thickness cryostat bottom	-	65 mm
LN_2-volume for 1100 mm filling level	-	125 l
cryostat material	-	glass fibre reinforced plastic (GFRP)
maximum operation pressure in insulation vacuum gap	-	$5 \cdot 10^{-4}$mbar
leakage rate	-	$1 \cdot 10^{-9}$mbar $\cdot$ l/s
measured heat intake at 1 m filling level (manufacturer)	$P_{\text{loss,cryo}}$	55 W
cryostat manufacturer	-	ILK Dresden GmbH

Due to the construction and insulation of the transformer cryostat, the bottom part of the LN_2-gap cannot be mechanically loaded with the weight of the secondary winding. Thus, the winding is suspended from the top of the transformer assembly.
The properties of the transformer cryostat are listed in Table 4.3.

Gas-Tight Cryostat Lid

A calorimetric measurement method was used for determining the AC-losses and the additional loss mechanisms in the secondary transformer winding. The measurement technique is described in Chapter 5.3.1.
The calorimetric measurement method relies on the continuous monitoring of the evaporation rate of the liquid nitrogen inside the cryostat, while the transformer is in operation at different load levels. This requires a gas-tight lid, closing off the cryostat and allowing the attachment of a flange for connecting a volumetric flow meter.

Figure 4.6: Sectional view of the transformer cryostat manufactured from non-conductive GFRP-material (computer aided visualization)

Since the maximum differential pressure across the volumetric flow meter is specified to be smaller than six millibar, the sealing of the cryostat is done in a simple manner. For this purpose the cryostat lid is made of ten millimeter thick PVC-material and sealed with circular seals made of flexible silicone tubing. The silicon seals are placed in grooves cut into the PVC-lid and are additionally lubricated with high vacuum grease to improve the leak tightness during thermal cycling. Due to the relatively low pressure inside the cryostat the lid can simply be kept in place by the use of tight-fitted Styrofoam® spacers between the top of the lid and the iron core clamping bars.
The current-leads to the secondary winding are also fed through the cryostat lid without allowing any gaseous nitrogen to escape from the cryostat. This is done using thin-walled and flexible silicone hoses with a diameter of 40 mm. At the bottom, the hoses are connected to flanges in the cryostat lid. At the top end, the silicon hoses are joint to the current leads using gas-tight threaded bushings around the brass rods of the current leads. This creates a gas-tight connection between the cryostat lid and the current leads without connecting the two parts mechanically.
The flexibility of this connection is important, since the cryostat as well as the cryostat lid are mechanically connected to the bottom part of the iron core, whereas the superconducting winding together with the current leads are suspended from the iron core yoke at the top of the transformer assembly. This required a flexible connection to compensate for any thermal contractions of other small movements in the whole assembly. Figure 4.7 shows a visualization of the cryostat-lid with the silicone sealing sitting on the LN_2-gap of the cryostat. The cryostat-lid further prevents moisture from the ambient air to condensate inside the cryogenic environment of the transformer setup.
During the short-circuit current limitation measurements the volumetric flow of gaseous nitrogen is not measured and therefore the cryostat does not need to be sealed gas-tight. For the limitation measurements, the cryostat- lid serves purely as a moisture barrier for the cryogenic environment and is sitting loosely on top of the cryostat. This allows it to open automatically for releasing the higher pressure that is build up during the quench of the superconductors in the secondary winding and the consequent rapid evaporation of LN_2.

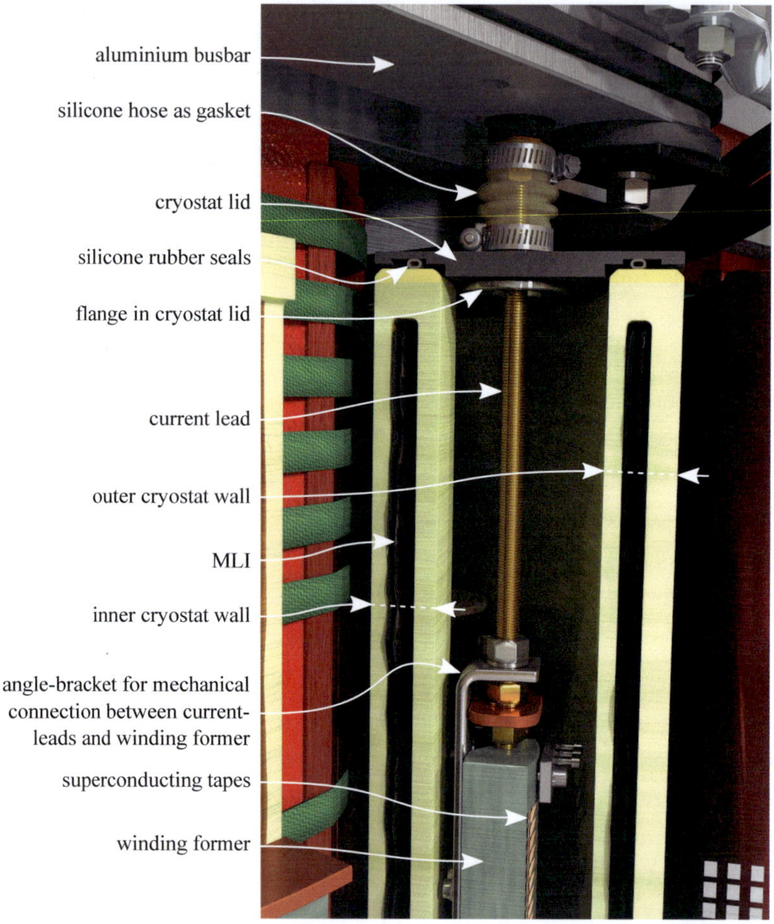

Figure 4.7: Optimized brass current lead, connecting the aluminium busbars, at room-temperature level, to the copper busbars inside the transformer cryostat. The current leads also provide the mechanical suspension for the secondary winding (computer aided visualization)

4.5 Superconducting Secondary Winding

The construction of the superconducting secondary winding is divided into three main parts. These are the current leads, the winding former and the superconducting winding itself. The three main parts of the superconducting winding are explained in the following.

Figure 4.8 shows the full assembly of the superconducting winding.

4.5.1 Current Leads

In the transformer design the current leads are connecting the superconducting winding inside the cryostat to the transformer load outside the cryogenic environment at room-temperature level. In this setup the current leads serve two different purposes.

The first purpose is the efficient current transport from room-temperature level to 77.3 K inside the cryostat with a minimal heat intake due to thermal conduction and due to Joule-heating.

The second purpose of the current leads is the mechanical suspension of the whole secondary winding assembly. This requires that the current leads are not only electrically coupled to the superconducting tapes inside the cryostat but also mechanically connected to the structure of the winding former. Outside of the cryostat the current leads are electrically and mechanically connected to the aluminium busbars which are in turn suspended from the iron core yoke clamping bars.

The current leads were numerically optimized to find the ideal diameter using an approach based on [Wil83, See98, CL12]. The length of the current leads and the range of the LN_2-filing level is given by the transformer design as discussed in Chapter 2.3. The current leads are made of brass with a copper to zinc ration of 65 % to 35 % in material weight. This composition is also referred to as *high brass* and it has a lower conductivity and a higher tensile strength compared to standard copper. This means that an optimized current lead made of brass has a about four times larger cross section area compared to a optimized current lead manufactured from copper. A brass current lead can carry more mechanical load due to its higher tensile strength while, providing the same thermal properties as a copper current lead. This makes brass an appropriate choice for the application as an optimized current lead in combination with the required mechanical suspension for the superconducting winding assembly.

To guarantee the symmetry of the mechanical suspension and a symmetric magnetic behavior of the secondary winding, a total of six current leads is used instead of only two - three for the inward-current and three for the outward-current. The current leads are

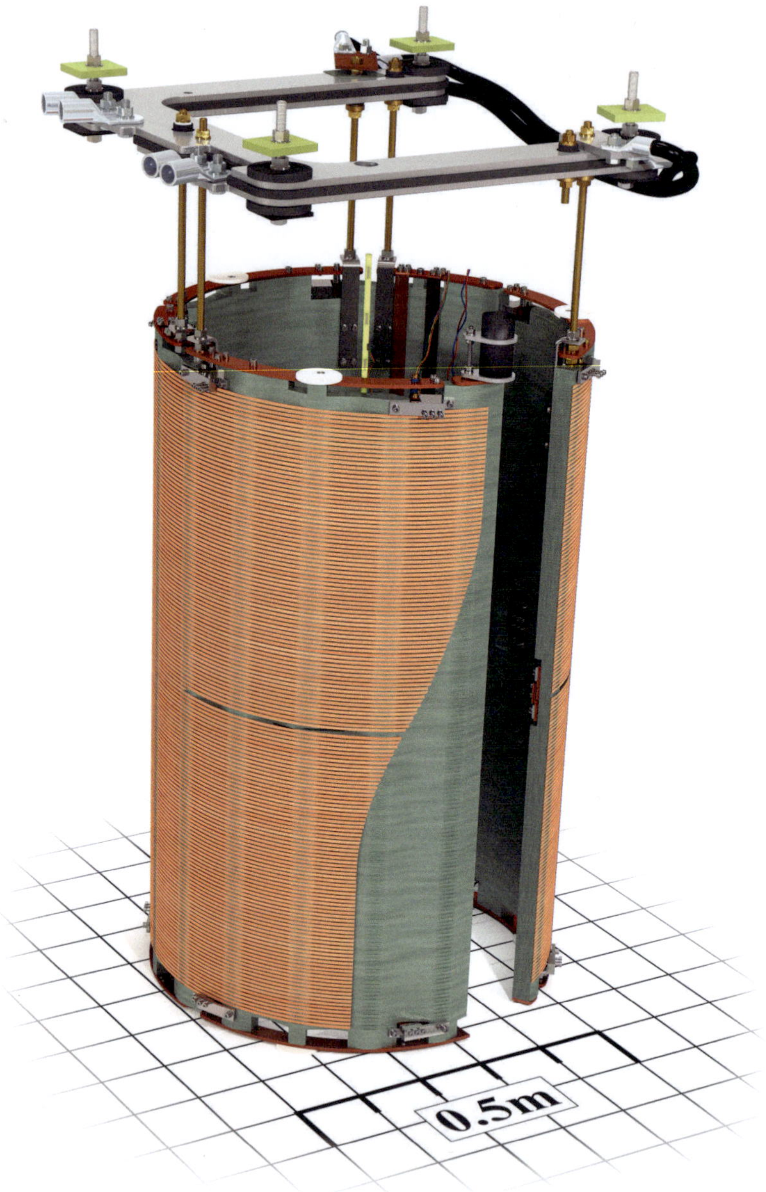

Figure 4.8: The superconducting secondary winding of the transformer as a cutaway view. The visualization also shows the brass current leads, the aluminium busbars, the copper busbars and the contacts, terminating the six superconductor stacks (computer aided visualization)

manufactured of threaded brass rods with a cross section area of 55 mm^2. This corresponds to a M10 - ISO metric thread according to the standard DIN 13. At the optimum point of operation each current lead carries a current of 192.5 A (RMS), which is one third of the total nominal secondary current.
Figure 4.7 shows a visualization of the current lead assembly based on computer aided design drawings.

4.5.2 Glass Fibre Reinforced Plastic Winding Former

Compared to a conventional copper winding, carrying a nominal current of 577.4 A, the superconducting winding has a significantly reduced cross section area. Therefore, the superconducting tapes have a significantly reduced mechanical stiffness and cannot be arranged in a self-supporting manner to mechanically withstand the magnetic forces during a fault current. For this reason the design utilizes a non-conductive GFRP-winding former to keep the superconducting tapes in place and to mechanically support the stacked tapes during a fault current event as well as during the nominal operation.
The winding former is machined out of a single piece of a GFRP-tube with an original wall thickness of 30 mm and an inner diameter of 540 mm. To allow for the thermal expansion of the superconducting material during a quench and thus to avoid radial buckling of the tapes [LCEB11] the outside surface of the winding former is not of a cylindrical shape but has a repetitive pattern of vertical grooves along the circumference.

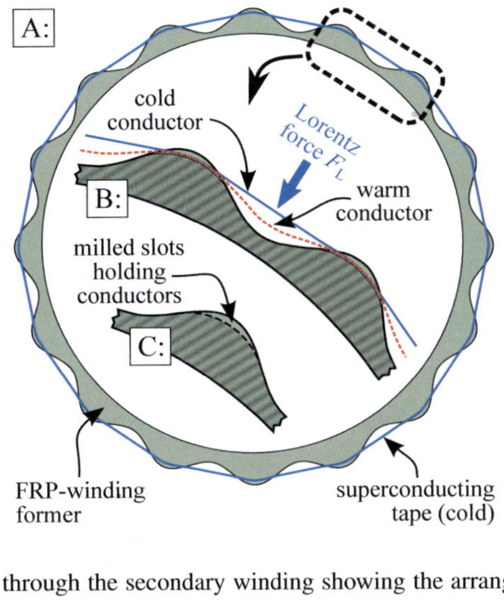

Figure 4.9: Horizontal cut through the secondary winding showing the arrangement of the superconducting stacks in normal operation at 77.3 K (blue) and in quenched condition with T_{sc} well above 77.3 K (red) (drawing is not to scale)

Table 4.4: Properties of non-conductive GFRP-winding former

property	value
winding former height	964.5 mm
outer radius	300 mm
inner radius	270 mm
minimum winding former wall thickness (grooves)	20 mm
maximum winding former wall thickness (raisings)	30 mm
number of vertical grooves along circumference	18
winding former weight (manufacturer)	~120 kg
width of horizontal slots for tape stacks	4.5 mm
spacer thickness between horizontal slots	1.5 mm
winding former manufacturer	Fabrum Solutions Limited

A sketch presenting a horizontal cut through the winding former is shown in part A of Figure 4.9 with the superconducting tape being depicted as a blue continuous line. Part B of Figure 4.9 shows the comparison of the tape positioning during the normal operation as a blue line and the position of the expanded tape during a quench, represented by a red dashed line. During a fault the expanded tape is pressed into the vertical grooves by the Lorentz-force F_L, which the superconductor experiences due to the magnetic stray field between both windings. The outer radii of the winding former, that the superconducting tape is in contact with, are chosen to be well above the minimum bending radius of the HTS-material [ZYS+15] to prevent mechanical damage in all operation conditions.
This novel winding former design allows efficient protection of the tapes against radial buckling and also offers direct contact with the coolant on both sides and therefore better cooling of the tape-stacks. The winding former design was filed for patenting under the application number PCT/EP2016/001412 and DE 102015110442 A1.
Additional slots, following the helical path of the conductors in the winding arrangement were machined into the raisings in the outer contour of the winding former in order to keep the tape arrangements at their destined vertical positions. A closeup of the vertical grooves and the the helical slots in the winding former is shown in the photography in Figure 4.11 and in Figure 4.12 in Chapter 4.5.3.
Table 4.4 lists the properties of the GFRP-winding former. A computer aided visualization of the winding former is shown in Figure 4.10.

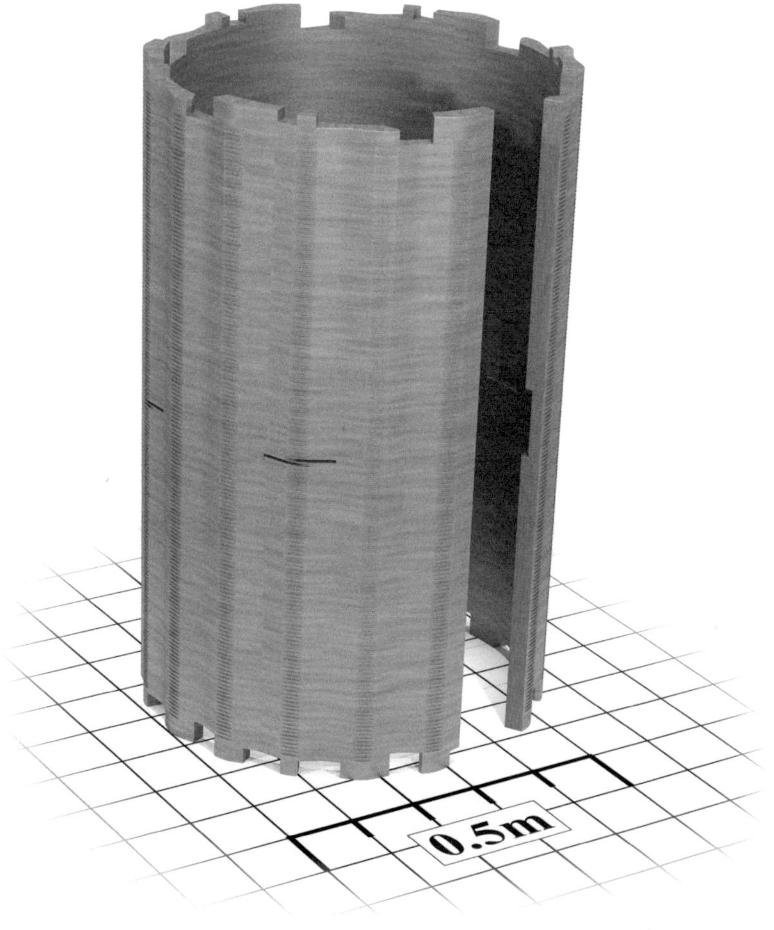

Figure 4.10: The non-conductive winding former machined out of a GFRP-tube. The "horizontal" slots along the circumference of the winding former hold the stacks of superconducting tapes in position (computer aided visualization)

4.5.3 Superconducting Winding

The superconducting winding is wound directly in the machined slots of the winding former. The winding consists of twelve parallel tapes with an average critical current of 160 A per tape. The superconducting tapes are arranged in six stacks of two tapes each, as discussed in Chapter 2.3. To allow for current distribution measurements between the two tapes in one stack, the tapes are insulated to each other by interlacing two 50 μm Kapton® films between the superconductors. The detailed tape arrangement in the superconducting winding is shown in Figure 2.18 in Chapter 2.

In order to allow separate current measurements, the tapes in the stacks are separately connected via pressure contacts on the top and bottom end of the winding. To keep the whole magnetic arrangement of both, the primary and the secondary winding as symmetric as possible the six contacts terminating the superconducting stacks are distributed along the upper and lower rim of the winding former. This results in fully symmetric positions with an angular distance of 60° between the six contacts along the circumference of the winding former. The magnetic symmetry is important to keep any electric compensation current between the six parallel stacks as low as possible.

For the manufacturing of the winding, the two superconducting tapes of one stack and the two Kapton® films are wound simultaneously with a controlled tension of 60 N per superconducting tape and a tension of 5 N for each of the Kapton® films respectively. The winding process shown in a photography in Figure 4.11. The high tension in the single superconducting tapes compensates for the larger thermal contraction of the GFRP-winding former in comparison to the superconductor material after cooling from room-temperature to operating conditions at 77.3 K.

Figure 4.12 shows the secondary winding while being lifted into the operating position inside the transformer cryostat and the primary winding.

A list of properties for the secondary winding is given in Table 4.5.

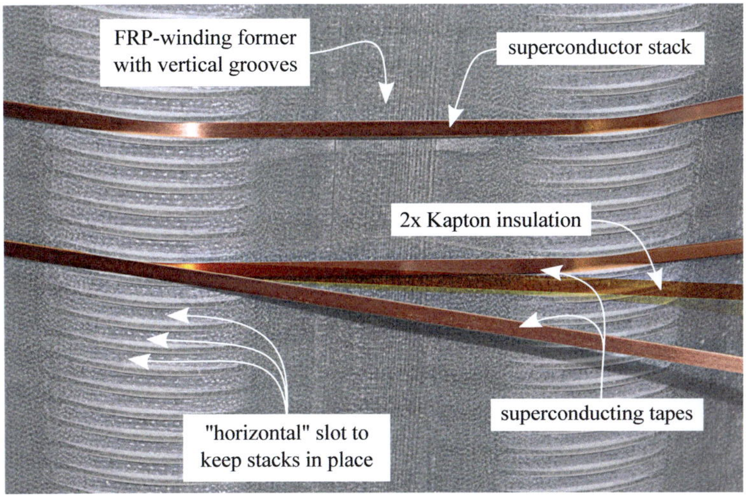

Figure 4.11: The GFRP-winding former during the winding process. Two superconductors and two Kapton® insulation films are wound simultaneously into one stack

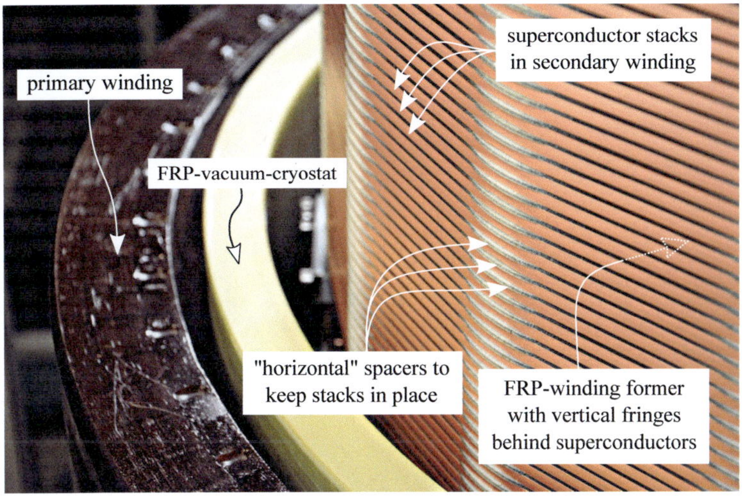

Figure 4.12: Superconducting secondary winding while inserting into the GFRP-cryostat and the primary winding

Table 4.5: Properties of superconducting secondary winding

property	symbol	value
nominal operation current	I_s	577.4 A
nominal operation voltage	U_s	1000 V
turn-to-turn voltage	u_W	40 V
parallel conductor configuration	-	6 x stack of 2
number of parallel superconducting tapes	$n_{s,par.}$	12
arrangement of clamping contacts (normal conductor $\leftrightarrow$ superconductor)	-	with 60° spacing around circumference on top and bottom rim
number of turns in winding	N_s	25
length of single tape	-	48.81 m
total superconductor length	-	561.72 m
total winding weight (including contacts, current leads, busbars, winding former and superconductors)	-	~150 kg
radial tension per superconductor tape	-	60 N
radial tension per Kapton® insulation tape	-	5 N

5 Experimental Setup for Testing a 1 MVA-Class Current Limiting Transformer

This chapter describes the experimental setups used for the different measurements on the superconducting current limiting 1 MVA-class transformer demonstrator. The manufacturing and the assembly of the components of the transformer demonstrator is described in Chapter 4. In addition, this chapter describes the methodology and the measurement techniques that are applied for the data collection and the generation of the measurement results. This explanation is mandatory to understand the results of the measurements presented in Chapter 6.

Firstly, the measurement on the temperatures of the multi-layer insulation inside the vacuum cryostat is described in section 5.1. Chapter 5.2 explains the experimental setup for determining the fundamental parameters of the 1 MVA-class transformer demonstrator. These parameters are the short-circuit impedance u_k, the current-sharing between the parallel superconducting tapes of the secondary winding and the measurement on the iron core magnetization. The results of the magnetization measurement are used to determine the iron core losses, as a part of the transformer loss analysis, which is described in Chapter 5.3.

Further, Chapter 5.3 presents two experimental procedures to determine the losses inside the transformer cryostat as well as the experiments conducted to measure the transformer losses for different transformer loads.

Lastly, Chapter 5.4 shows the experiments for testing the fault current limitation capabilities of the 1 MVA-class transformer demonstrator.

All devices used for the measurements on the transformer demonstrator mentioned in this chapter are listed in Table 8.2 and Table 8.3 in Chapter 8.4 of the appendix. The explanation of the experiments in the following sections will refer to the devices (D1 to D19) in Table 8.2 and Table 8.3.

5.1 Multi-Layer Insulation Temperature Measurement

The construction of the transformer cryostat including the multi-layer insulation (MLI) is explained in Chapter 4.4. The use of non-conductive materials for the cryostat construction is necessary to avoid induced eddy-currents. This applies especially for the outer wall of the cryostat, since it is operated in the magnetic AC-field of the stray-gap between both transformer windings.
The cryostat material is made from non-conductive GFRP-material, whereas the thermally reflective sheets of the multi-layer insulation are aluminium coated and hence electrically conductive. If the multi-layer insulation is operated in a transient magnetic field, induced voltages and eddy-currents in the thin conductive layers of the MLI cannot be avoided. Eddy-currents can heat the material of the MLI and significantly reduce the efficiency of the thermal insulation of the cryostat.
To evaluate the applicability of standard, non-segmented MLI in AC-magnetic fields of a superconducting transformer, temperature measurements at specific points on the MLI are performed, during no-load and nominal-load operation of the transformer.
Figure 5.1 shows the positions of eight temperature sensors (D1), applied in the MLI of the transformer cryostat. The drawing additionally gives a visualization of the numerically calculated magnetic flux density in the transformer. The magnetic flux density is shown in normalized absolute values.
The positions for the temperature sensors are selected for points with high perpendicular magnetic fields, as the heating effect due to the eddy-currents at those positions is expected to be most significant (T1, T2, T3, T4, T5, T7 and T8). As a reference measurement point, one temperature sensor (T6) is positioned in the middle of the stray-gap. At this position, only magnetic fields parallel to the MLI and thus no eddy-currents are expected. The temperature sensor T3 is mounted between both stacks of MLI-foils and sensor T4 is mounted on the inside of the MLI, facing the cold surface of the cryostat wall. All other temperature sensors are fixed to the outside of the MLI-foil stacks.
For the acquisition of the MLI temperatures, the resistive elements of the sensors are connected in series and supplied with a 1 mA measurement current using a precision current supply (D2). The readout of the measured values is done with a universal data acquisition device, (D3) to which the temperature sensors are connected directly, using a four-lead connection technique and a differential input configuration.
The temperature measurement is conducted during the nominal load measurements, as part of the loss-measurements, described in Chapter 5.3.2.

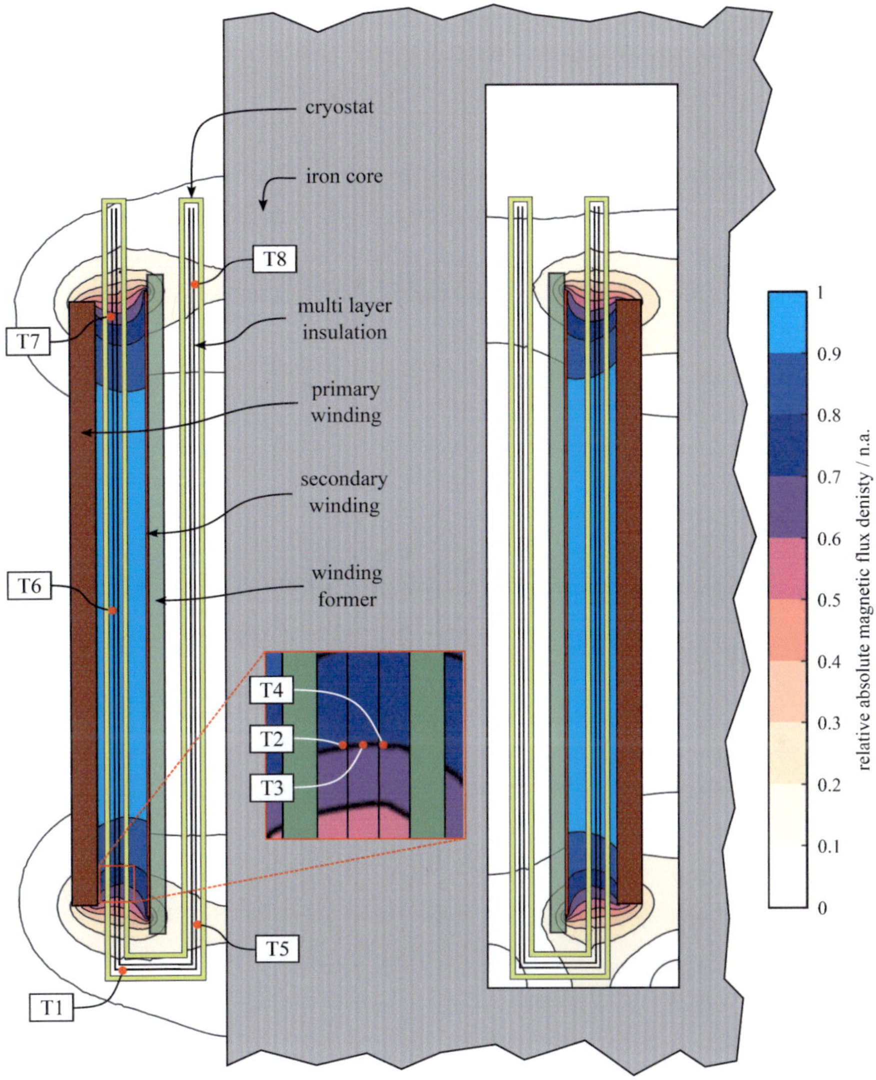

Figure 5.1: The drawing shows the positions of the temperature sensors T1 to T8 in the multi-layer insulation with respect to the simulated absolute magnetic flux density. All sensors but sensor T3 and sensor T4 are mounted on the outer side (facing the room-temperature side of the cryostat wall) of the MLI-foil stacks (drawing is not to scale)

5.2 Measurement of Fundamental Transformer Parameters

This chapter discusses the measurements of the short-circuit impedance, the current-sharing in the secondary winding as well as the magnetization of the iron core of the transformer demonstrator.

The measurement of the short-circuit impedance, shown in Chapter 5.2.1, was conducted following the same procedure as applied for conventional transformers, using a reduced voltage on the primary transformer winding and a short-circuited secondary winding. However, the measurements of the current-sharing in the secondary winding and of the iron core magnetization, explained in Chapter 5.2.2 and Chapter 5.2.3, were conducted while operating the transformer at rated voltage- and current values.

5.2.1 Short-Circuit Impedance Measurement

The short-circuit impedance measurement is part of the standard type-tests conducted on transformers in accordance with IEC 60076-1 [CFSW03, Geo09]. It is used to evaluate and to compare the magnetic properties of the transformer assembly and thus the behavior of the transformer in a power grid. The measurement procedure is identical with the procedure as it would be conducted on a conventional transformer.

For determining the short-circuit impedance, u_k, the transformer taps of the secondary winding are short-circuited. The primary transformer voltage is then gradually increased while the short-circuit current through the secondary winding is monitored. The voltage, applied to the primary transformer winding, that causes the nominal operation current to flow in the secondary winding, is called the short-circuit voltage U_k. With the short-circuit voltage on the primary taps and the current in the primary winding, the impedance value of the short-circuited transformer, Z_k can be calculated with the following Equation 5.1.

$$Z_\mathrm{k} = \frac{U_\mathrm{k}}{I_\mathrm{p}} \tag{5.1}$$

The measured short-circuit voltage U_k can be transferred into to the short-circuit impedance u_k with Equation 5.2.

$$u_\mathrm{k} = \frac{U_\mathrm{k} \cdot 100\,\%}{U_\mathrm{r}} \tag{5.2}$$

The short-circuit impedance, u_k is the percent ration between the short-circuit voltage and the rated primary voltage, U_r of the transformer. In the literature, the short-circuit

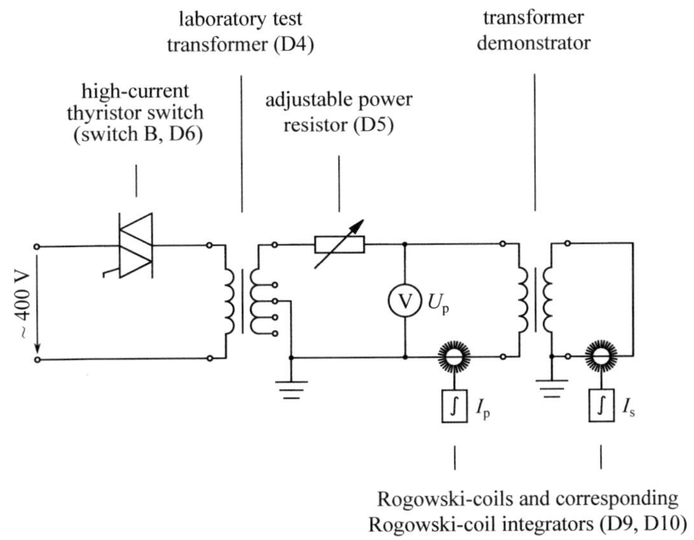

Figure 5.2: Circuit-diagram for the experimental setup to measure the short-circuit impedance of the transformer demonstrator

impedance, u_k is also referred to as *relative short-circuit* voltage or *impedance voltage* [Geo09].

The measurement circuit for the short-circuit test is depicted in Figure 5.2.

The transformer demonstrator is shown on the right hand side of the circuit-diagram. The primary voltage is supplied by a laboratory test transformer (D4) with an adjustable secondary voltage, that can be increased, starting from 50 V up to 1500 V in steps of 50 V. An adjustable power resistor (D5) is used in series to the laboratory test transformer for the fine-tuning of the primary voltage on the transformer demonstrator.

The current in the measurement circuit is switched to the transformer demonstrator using a bipolar thyristor switch (D6), which is triggered by the main measurement computer. The thyristor switch allows precise timing for the duration of the short-circuit testing and a simple synchronization with the data acquisition equipment.

During the measurement, the primary voltage of the transformer demonstrator as well as the primary and secondary current are acquired and stored using a transient recorder (D7).

For measuring the primary transformer voltage the primary winding taps are connected to the inputs of the transient recorder via a differential probe (D8) with an adjustable signal attenuation. For the primary- and secondary current measurements, Rogowski-coils (D9) are used in combination with a multi-channel Rogowski-coil integrator (D10). The output signals of the Rogowski-integrators are acquired and recorded by the transient recorder.

5.2.2 Current-Sharing between Superconducting Tapes

The superconducting secondary winding of the transformer is composed of twelve parallel superconducting tapes, which are arranged in six stacks of two tapes each. As described in Chapter 4.5, the contacts, terminating the tape stacks are distributed along the circumference of the winding former in a magnetically symmetric manner with a angular distance of 60°.
To test the magnetically symmetric design and the current distribution between the six tape stacks, a measurement is conducted, in which the current in all superconducting tapes is measured individually while the transformer demonstrator is operated under full load and at nominal conditions.
To measure the current in the parallel superconducting tapes, twelve custom-made, liquid nitrogen immersible Rogowski-coils (D11) were designed and manufactured. The winding-former in the Rogowski-coils, holding the coil-wire in place, is manufactured from 3d-printed acrylonitrile-butadiene-styrene, also known as ABS. The design of the Rogowski-coil former provides well defined grooves to hold the coil-wire in place and to ensure symmetrical distribution of the wire around the coil. This is important to guarantee a homogeneously wound coil and consequently a coil sensitivity that is independent from the positioning of the coil around the path of the measured current.
A photography of two Rogowski-coils in the secondary winding assembly is shown in Figure 5.3. The coils are inserted around the current path of each superconducting tape, between the copper busbars and the pressure contacts inside the transformer cryostat.
For the current-sharing measurement, the transformer demonstrator is operated at the rated primary voltage of 20 kV and loaded with the nominal resistive load of 1.73 Ω. Figure 5.4 shows the circuit diagram of the experimental setup for measuring the current-sharing between the conductors in the secondary winding.
The transformer demonstrator is connected between two phases of the 20 kV medium-voltage grid via a high-voltage thyristor switch (D12). The thyristor switch is triggered by the main measurement computer and allows precise and save connecting and disconnecting of the measurement setup to and from the 20 kV-grid. During the measurement, the transformer is loaded with a customized load resistor (D13) with a resistance value of 1.73 Ω. The load resistor is designed for a maximum voltage of 1500 V and is actively cooled in order to withstand a nominal power of up to 600 kW for a maximum duration of 120 s.
The twelve Rogowski-coils in the secondary winding assembly are connected to custom-

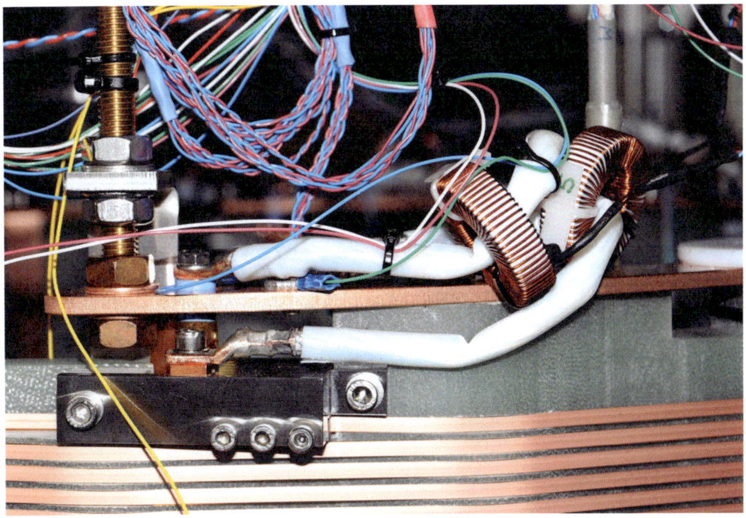

Figure 5.3: Photography of custom made Rogowski-coils allowing the immersion in liquid nitrogen and therefore the use inside the transformer cryostat

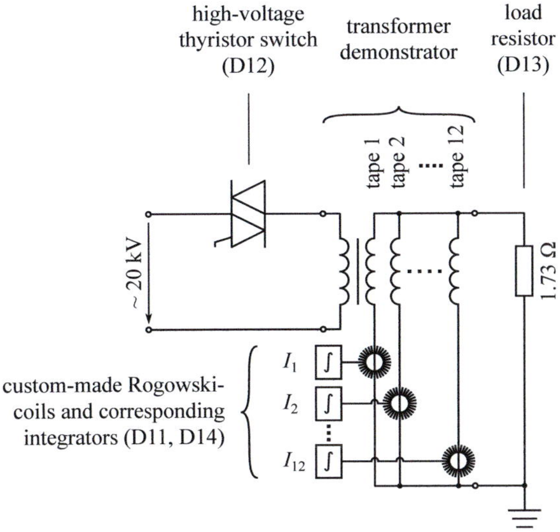

Figure 5.4: Circuit-diagram for the experimental setup to measure the current-sharing between the superconducting tapes in the secondary winding of the transformer demonstrator

made Rogowski-coil integrators (D14). The integrators are adjusted to the resistance and inductance of the connected Rogowski-coils to ensure a linear response of the current measurement over a wide range of frequencies. The transient output signal of the Rogowski-coil integrators is assessed with a universal data acquisition device (D15). The Rogowski-integrators are connected in a single-ended configuration and the start of the data acquisition is synchronized to the trigger signal of the high-voltage thyristor switch.

5.2.3 Iron Core Magnetization and No-Load Parameters

The measurement setup of the iron core magnetization of the transformer demonstrator consists of two pick-up coils. The coils are used to determine the magnetic flux density in the iron core and therefore the saturation state of the material during a no-load measurement on the transformer demonstrator. Since the iron core magnetization does not depend on the actual transformer load but solely on the applied primary voltage, the acquired results apply also for the operation of the transformer under nominal load.

The core magnetization is used to calculate the iron core losses, as a part of the transformer loss analysis described in Chapter 6.3.3, as well as to verify the magnetic flux density value specified during the transformer design process described in Chapter 2.3.

The pick-up coils are made of five turns of litz-wire contained inside a flexible plastic tube. One of the two pick-up coils is arranged around the unwound limb of the transformer, as it is shown in Figure 5.5. The coil is positioned equidistant between the top- and the bottom yoke of the transformer iron core.

The second coil is placed inside the stray-gap of the transformer at the same height as the first coil. The coil in the stray-gap surrounds not only the iron core limb of the transformer but also the secondary transformer winding together with the transformer cryostat.

The peak value for the pick-up voltage is calculated using Equation 5.3. For a turn-to-turn voltage, u_{w} of 40 V and the number of turn in the pick-up coil, $n_{\mathrm{pick-up}}$ being 5, this results in an amplitude of 282.8 V for a sinusoidal change in the magnetic flux with a frequency of 50 Hz.

$$U_{\mathrm{pick-up}} = \sqrt{2}\, n_{\mathrm{pick-up}}\, u_{\mathrm{w}} \tag{5.3}$$

For the voltage measurements on the pick-up coils, an universal data acquisition device (D15) is used. Since the expected pick-up voltage during no-load operation of the transformer exceeds the maximum input voltage of the acquisition device, the coils are connected via a differential probe (D8) with an adjustable attenuation-ratio.

The measurement of the transformer iron core magnetization is conducted using the mea-

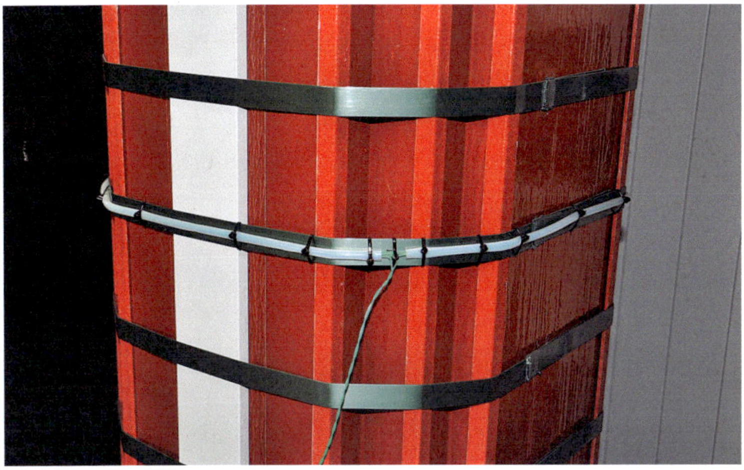

Figure 5.5: Pick-up coil around the limb of the transformer iron core. The coil is placed in the middle between the top and the bottom core yoke

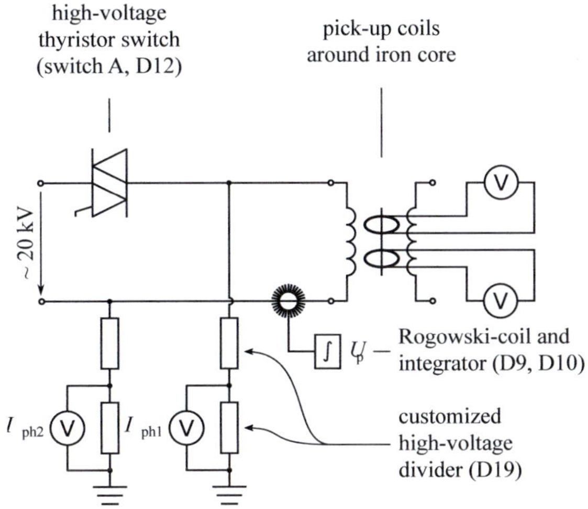

Figure 5.6: Circuit-diagram for the experimental measurement setup to determine the iron core magnetization and the no-load parameters of the transformer demonstrator

surement circuit shown in Figure 5.6. The iron core magnetization depends on the primary transformer voltage and thus, requires the rated voltage on the primary winding of the transformer demonstrator. During the measurement the transformer is switched to two phases of the 20 kV-grid, using a bipolar thyristor switch (D12), while the transformer secondary taps are open and the transformer is in no-load operation. The data acquisition for the measurement is started about ten seconds after the transformer demonstrator is switched to the 20 kV-grid, to ensure that all transient oscillation has seized.
The phase-to-phase voltage of the 20 kV grid cannot be directly measured via a voltage divider. Therefore the voltage of both phases is determined with respect to ground potential using two customized high-voltage dividers (D19). The transformer primary voltage can be calculated as the difference between U_{ph1} and U_{ph2}.
The primary current during no-load operation is measured with a Rogowski-coil (D9) and a corresponding Rogowski-coil integrator (D10).

5.3 Transformer Loss Measurement

The transformer loss measurements can be separated into three main contributing parts: the measurement of the primary winding losses, the measurement of the iron core losses, and the loss measurement inside the cryogenic environment of the transformer. The loss contributions are described in Chapter 2.2.1.
The primary winding losses $P_{\mathrm{loss,p}}$ are caused by Joule-heating in the conductor material. They can be calculated from the pre-determined resistance of the primary winding, R_{p} as specified in Chapter 4.3, and from the measured primary current I_{p}. The dielectric losses inside the insulation material of the primary winding are neglected for this loss analysis.
The losses in the iron core are obtained from the magnetic flux density in the iron core in combination with the specific core material losses $P_{\mathrm{Fe,spec.}}$, determined by the manufacturer of the iron core material and listed in Chapter 4.2. The measurement of the flux density in the iron core is explained in Chapter 5.2.3. With the measured value of the magnetic flux density, the losses in the iron core, $P_{\mathrm{loss,Fe}}$ can be calculated.
The nominal magnetic flux density in the iron core is only reached for a transformer operation at nominal primary voltage. This makes it necessary to conduct the magnetization measurement on the iron core at the rated 20 kV primary voltage. The measurements of the remaining loss contributions require the nominal currents in the windings only and can be conducted at lower voltages. This significantly reduces the safety considerations for the experimental setup.
For determining and analyzing the losses in the cryogenic environment of the trans-

former, the measurement techniques described in Chapter 5.3.1 are applied.

5.3.1 Calorimetric Cryostat Measurements

The losses inside the cryostat of the transformer demonstrator originate from three different loss mechanisms. One part of the losses is caused by the Joule-heating in the contacts inside the secondary winding, in the busbars of the winding construction as well as in the current leads, connecting the secondary winding through the boundaries of the cryogenic environment. The second part is the heat flow through the cryostat walls and along the current leads, from room-temperature level to the temperature level of the liquid nitrogen. The third part are the AC-losses in the superconducting tapes of the secondary transformer winding.
The determination of these loss contributions via separate measurements would require a complex measurement setup, addressing each of the different loss mechanisms individually. Thus, two alternative calorimetric measurements were conducted during the transformer loss measurements in this project.
The calorimetric measurement allows the determination of all losses inside the transformer cryostat as a single loss value, that is derived from the amount of evaporated liquid nitrogen. The separation of the measured value for the total heat intake into its different loss components in done in a later step.
Both methods for measuring the losses inside the cryostat were used simultaneously during the transformer loss measurements and are explained in the following.

Loss Analysis via the Volumetric Flow of Gaseous Nitrogen

The experimental setup shown in Figure 5.7 is based on monitoring the volumetric flow of gaseous nitrogen leaving the transformer cryostat. In the following sections this measurement is referred to as *loss-measurement A*. The measurement method requires a gas-tight lid closing the transformer cryostat towards the environment, as explained in Chapter 4.4, to ensure that all evaporated LN_2 passes through the volumetric flow meter (D16). The volumetric flow meter additionally determines the temperature of the evaporated gas at the exhaust point into the surrounding atmosphere. From the volumetric flow value and the gas temperature it is possible to calculate the total rate of heat flow, $\dot{Q}_{\text{total}}$ into the cryogenic system.
For keeping the temperature of the volumetric flow meter above the minimum operating temperature of -10 °C, as specified by the manufacturer, the device is not directly connected to the exhaust port of the cryostat but separated by a low-temperature resistant hose. This provides an additional heat flow into the gaseous nitrogen, $\dot{Q}_{\text{hose}}$ along the

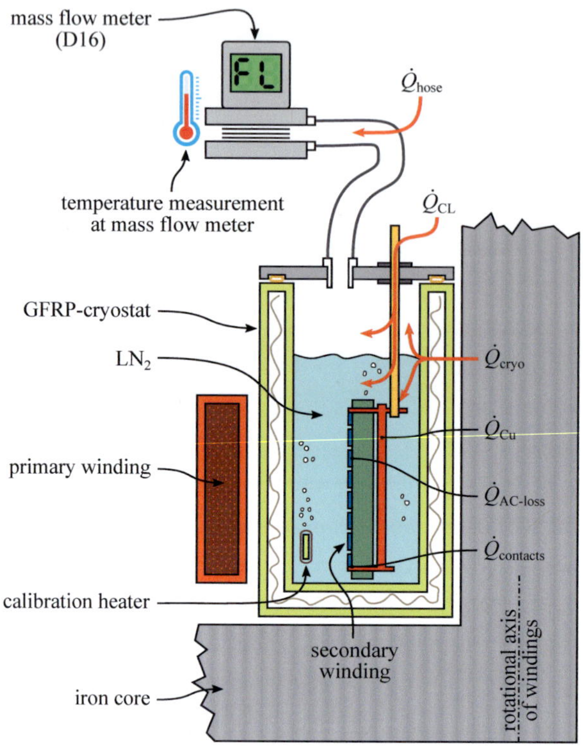

Figure 5.7: This cross section view of the transformer cryostat shows a volumetric flow meter, connected to the gas-tight cryostat, to measure the amount of evaporated liquid nitrogen. The sketch further depicts the different sources of heat flow into the cryogenic environment that influence the acquired loss value. This measurement is referred to as loss-measurement A (drawing is not to scale)

length of the hose and therefore prevents the temperature of the volumetric flow meter to drop underneath the specified minimum value during the measurements. In order to not significantly influence the gas flow through the volumetric flow meter with an increased build-up of differential pressure, the connecting hose is of a large diameter and the length of the hose is kept as short as possible.

The measurement setup does not allow to separate between the heat flow into the transformer cryostat itself and the heat flow into the hose, connecting the cryostat to the volumetric flow meter. For this reason, a further calibration of the measured total rate of heat flow into the cryostat is necessary. This is done by an additional calibration heater inside the transformer cryostat, as shown in Figure 5.7.

Figure 5.7 shows the separate heat flow contributions, which accumulate to the the total measured rate of heat flow, $\dot{Q}_{total}$.

During the measurement, the transformer is sequentially loaded for intervals of 60 min-

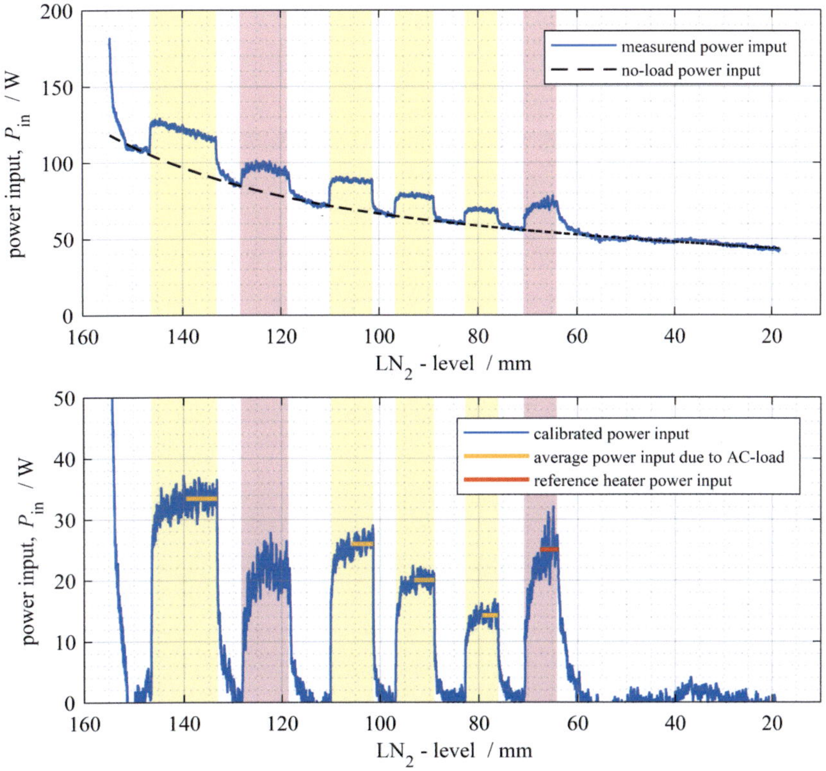

Figure 5.8: The top graph shows the total power input into the transformer cryostat as derived from the volumetric flow of gaseous nitrogen. The bottom graph depicts the power input into the cryostat after subtracting the no-load power input into the cryostat and after calibrating via the defined power input provided by the reference heater. AC-load phases are indicated by a yellow background color and the calibration with the reference heater is marked by a red background

utes, at different load levels, while the evaporation rate of the liquid nitrogen is continuously recorded. Figure 5.8 shows an example for a measurement sequence for determining the transformer losses during four different transformer load cycles. The transformer AC-load is indicated by the yellow background color in both graphs of Figure 5.8. The phases during which the setup is calibrated with the calibration heater in the cryostat are indicated by a red background color.

The top graph in Figure 5.8 shows the calculated power input into the cryostat (blue line) in Watt via the change in the LN_2-filling level in mm. The recorded values show an expected increase of the evaporation rate during the AC-load phases of the transformer as well as during the calibration heating phases. The dashed line in the top graph indicates the no-load losses versus the filling level of the full transformer assembly, as it is previously determined and averaged during multiple repeated no-load experiments. The

no-load losses include only the heat flow through the cryostat wall $\dot{Q}_{\mathrm{cryo}}$, the heat flow along the current leads $\dot{Q}_{\mathrm{CL}}$ and the warming-up of the gaseous nitrogen in the connecting hose $\dot{Q}_{\mathrm{hose}}$.
During the data analysis, the no-load losses are subtracted from the measured power input curve. The resulting curve can then be calibrated and scaled accordingly, using the defined power input into the cryostat that is provided during the reference heater intervals. The calibrated result is shown in the bottom graph of Figure 5.8.
It was found that the energy input during a AC-load phase of the transformer can be reliably acquired from the average value of the second half of the load or heater cycles. This average value is indicated by the yellow and red lines in the bottom graph of Figure 5.8. The values acquired during the first half of a load cycle show no consistent results, since the whole thermal system of the transformer requires a certain time to adjust to a electrical and thermal load change.

Loss Analysis via the Change in the Liquid Nitrogen Filling Level

Instead of monitoring the volumetric flow rate and the temperature of the exhaust gas, as done in loss-measurement A, the second measurement method relies on precise monitoring of the liquid nitrogen filling level inside the transformer cryostat. The change in the LN_2-filling level is then used to calculate the heat intake transformer cryostat from the amount of evaporated liquid nitrogen. In the following, this measurement method will be referred to as *loss-measurement B*.
The monitoring of the LN_2-filling level is done with a laser distance sensor (D17), measuring the distance between a fixed point outside the transformer cryostat and a buoyant Styrofoam® disk, floating on the surface of the liquid nitrogen. The Styrofoam® disk is held in place underneath the laser distance sensor with the help of a plastic guide-tube. The Styrofoam® disk has additional weights added to its lower side to increase its inertia and therefore reduce the noise in the measurement values.
For further smoothing of the measured LN_2-filling level values, the data acquisition is executed at a rate of 2.5 kS/s, while 2500 measurement values are averaged into one level-value per second. The averaging of the measurement values allows the elimination of the constant motion of the liquid nitrogen surface and the separation of the actual change in the LN_2-filling level.
The measurement setup for loss-measurement B is shown in Figure 5.9. For loss-measurement B it is not necessary to provide a gas-tight cryostat coverage. However, the sketch in Figure 5.9 depicts a lid with a transparent view-port on the transformer cryostat, because loss-measurement B was conducted in parallel to loss-measurement A, which requires a

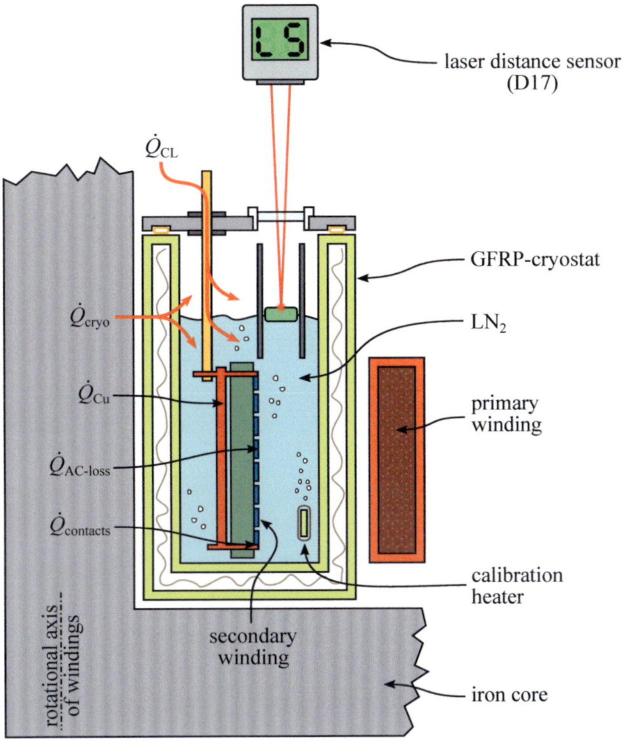

Figure 5.9: This cross section view shows the experimental setup for determining the total losses inside the transformer cryostat via the change in the liquid nitrogen filling level. The liquid nitrogen level is determined with a laser distance sensor positioned outside the transformer cryostat. This measurement is referred to as loss-measurement B (drawing is not to scale)

gas-tight transformer cryostat.

Figure 5.10 gives exemplary measurement results acquired with loss-measurement B, during the same experiment than the data presented for loss-measurement A in Figure 5.8. Phases with AC-load on the transformer demonstrator are indicated via a yellow background in the graphs whereas the calibration phases are indicated by a red background color.

The top graph in the figure shows the liquid nitrogen level inside the transformer cryostat versus the experiment time. The blue line is the actual measured result for the LN_2-filling level, influenced by the AC-transformer load, the calibration heater and the no-load heat flow into the cryostat. The black dashed line represents only the no-load level, influenced by the heat flow through the cryostat walls, $\dot{Q}_{cryo}$ and the heat flow along the current leads, $\dot{Q}_{CL}$. The no-load curve is previously acquired and averaged during multiple repeated no-load experiments.

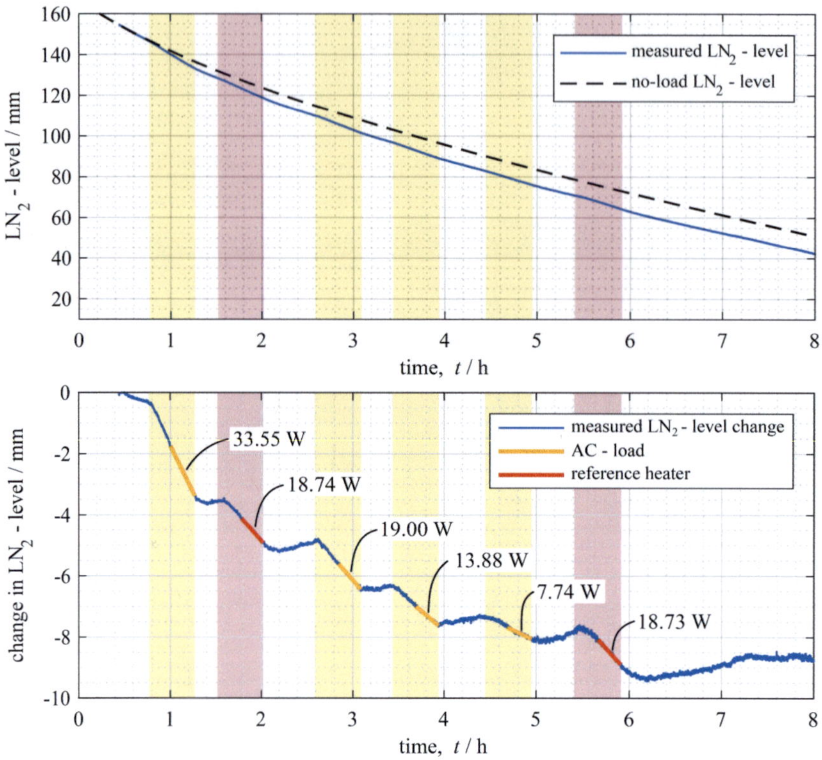

Figure 5.10: The top graph shows the total change in the LN_2-filling level during a AC-loss measurement including the change due to no-load losses. The bottom graph presents the change in the filling level without no-load power input into the cryostat and without heat flow along the current leads. AC-load phases are indicated by a yellow background color and the calibration with the reference heater is marked by a red colored background

To extract the heat intake, caused by the transformer load, the no-load LN_2-level (black dashed line in the top graph of Figure 5.10) is subtracted from the measured LN_2-level (blue line). The result is the change in the LN_2-level due to the AC-load of the transformer or the calibration heater, as shown in the bottom graph of Figure 5.10. The slope of the blue line represents the actual value for the heat flow into the cryogenic environment of the transformer during the measurement.

The heat intake value in Watt can be derived from the graph by linearly fitting the second half of the level change during a AC-load interval or a calibration heater interval. These linear fits, during the load intervals are represented by yellow and red lines in the bottom graph of Figure 5.10. As described for loss-measurement A, it was also found for the calorimetric measurement methods that averaging over the second half of the AC-load and calibration heater intervals yields the best results, since this allows the thermal system to stabilize after a load-change. The uncalibrated values for the power input, derived from the slopes in the change of the LN_2-level, are indicated in the bottom graph of Figure 5.10.

One significant advantage of loss-measurement B in comparison to loss-measurement A is the fact, that no gas-tight cryostat lid is required during the measurement. This allows the determination of cryogenic losses in a system also for cryostats that cannot be equipped with a gas-tight lid and also for simple experimental setups for which the construction of a gas-tight cryostat would be to intensive in cost.

The measurement method further allows the placement of the electronic laser sensor equipment to be far away from any magnetic or electric fields, without compromising the accuracy of the distance measurement.

The transformer loss results acquired with both described measurement methods, loss-measurement A and loss-measurement B, are presented in Chapter 6.3.2.

5.3.2 Loss Measurements at Variable Loads

A power transformer is usually most of the time operated well below the rated transformer load, during its life cycle in a power grid [Ber11, Geo09]. Therefore the analysis of the transformer performance at partial loads is important to estimate the transformer efficiency during different load conditions as they appear in a real power grid. For this reason, the measurements of the transformer losses, as described in this chapter, are not only evaluating the losses of the transformer at full load but also for loads smaller than the rated values.

The iron core losses are the only voltage dependent loss contribution. This means that all contributions to the total transformer losses, apart from the iron core losses, can be evaluated with nominal currents in the transformer windings but at lower voltage levels. This makes the experimental setup simpler in handling and reduces safety concerns, since no 20 kV grid connection is required. The iron core losses, presented as a part of the transformer loss measurement in Chapter 6.3.3 and Chapter 6.3.4, are obtained during the measurement on the current-sharing between the tapes of the secondary winding, as described in Chapter 5.2.2.

Figure 5.11 shows the circuit diagram of the experimental setup for the transformer loss measurement at variable load levels. The secondary winding of the transformer demonstrator is short-circuited, while the primary and secondary current are measured via Rogowski-coils (D9). The primary voltage of the transformer demonstrator is increased with a laboratory test transformer (D4) and an adjustable in-series resistor (D5), until the nominal current in the transformer windings is reached. This consequently causes nom-

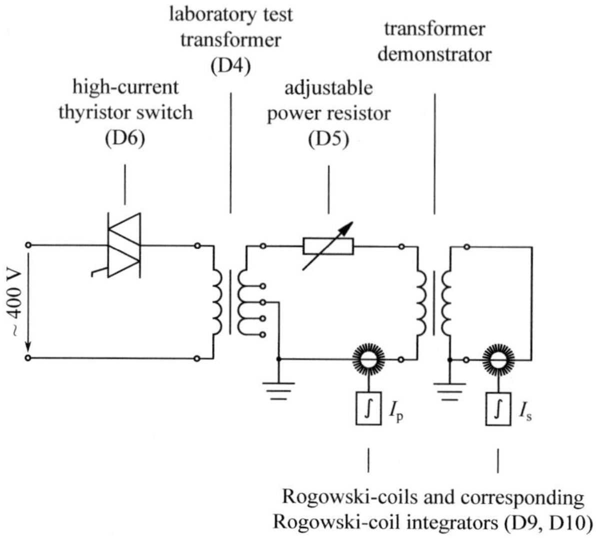

Figure 5.11: Circuit-diagram for the experimental setup to measure the losses of the transformer during variable transformer loads

inal transformer losses in the primary winding, the superconducting secondary winding and inside the transformer cryostat. Only the iron core losses are below the nominal value since the magnetization of the iron core scales with the primary voltage.

The primary winding losses can be determined from the current in the primary winding I_p and the primary winding resistance R_p. The losses in the superconducting secondary winding and in the transformer cryostat are measured with the calorimetric measurements methods, described in Chapter 5.3.1.

For the transformer losses at partial loads, the primary winding voltage is decreased in steps, while the measurements of the primary winding losses, the superconducting secondary winding as well as of the transformer cryostat are repeated for each load level. Since the thermal masses in the transformer cryostat and of the secondary winding assembly and are large and experience only a small additional heat intake due to the AC-load, the load intervals need to be of a certain length to allow the thermal system to adjust to a change in the load level. For the measurements presented in Chapter 6.3.2 and Chapter 6.3.3, it was found that the length of 60 minutes for one specific load level results in reproducible results. After measuring the transformer losses at one load level, the measurement setup is given sufficient time to recover back to the no-load loss level.

The data during the load loss measurement are recorded with a low acquisition speed of 1 S/s for the RMS-values of the primary and secondary current as well as for the volumetric flow rate of the evaporated liquid nitrogen, the nitrogen temperature and the liquid nitrogen filling level. The RMS-values of the primary and secondary current were determined using a universal data acquisition device (D15), operating at a sampling rate of 100 kS/s. The data acquisition device is acquiring the data and generating the RMS-values for intervals with a length of one second.

5.4 Current Limitation Measurement

The current limitation measurement described in the following is conducted to compare the measured values for the current limitation and the consequent temperature increase of the superconducting secondary winding to the design values and to the results of the current limitation prediction of Chapter 2.3. This comparison is important for refining the transient simulation process for the current limitation behavior of superconducting transformers and for optimizing the transformer design tools.
To measure the efficiency of the current limitation, the transformer demonstrator is integrated in a measurement circuit as shown in Figure 5.12.
At the start of the experimental sequence, the transformer is connected to the 20 kV-grid via a high-voltage bipolar thyristor switch (D12 / switch A). During the full experimental sequence the transformer is loaded with the nominal load resistor R_{load} of 1.73 Ω. This means the transformer is instantly switched into nominal operation.
After connecting the transformer demonstrator to the 20 kV-grid, a waiting time is required in order to allow for the transformer to reach a steady state operation. For the measurement results presented in Chapter 6.4 the steady state operation was reached in less than 200 ms, witch synchronizing the trigger point of switch A to 90° after the zero-crossing of the voltage in the 20 kV-grid.
The fault is produced by short-circuiting the load resistor with a high-current thyristor switch (D6, switch B) and an additional small in-series short-circuit resistor (D18).
To allow direct comparison between the simulated current limitation prediction and the measurement results, the trigger point of the high-current thyristor switch B is synchronized to the zero-crossing of the secondary current of the transformer demonstrator. Therefore, it is possible to start the current limitation phase exactly at the zero-crossing of the current.
The mains connection to the 20 kV power grid used during the experiments is limited to a maximum current of 900 A for a duration of 100 ms. For this reason the short-circuit of the transformer secondary winding is produced with a defined short-circuit resistor R_{short} instead of by fully short-circuiting the the load resistor. The short-circuit resistor in the experiments has a value of 25.3 mΩ, which leads, according to the current

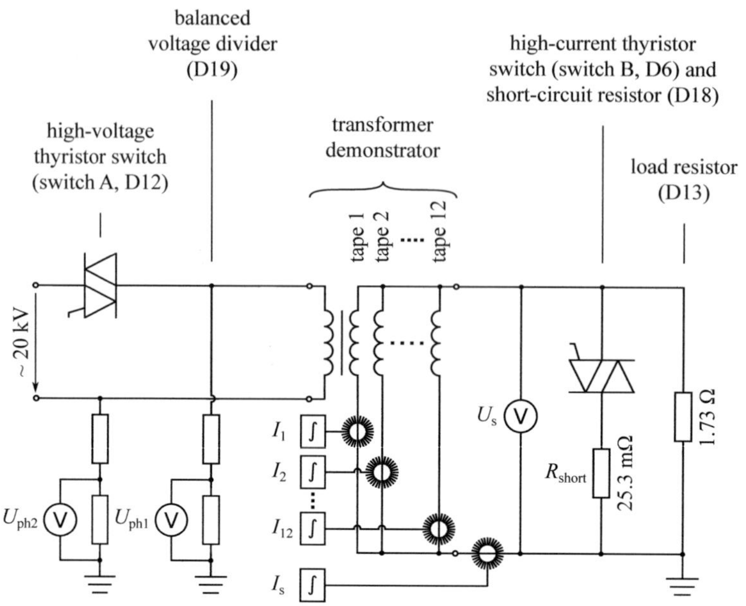

Figure 5.12: Circuit-diagram of the experimental setup to measure the current limitation behavior of the transformer demonstrator in combination with the current distribution between the parallel superconducting tapes of the secondary winding

limitation prediction from Chapter 2.3.3, to a maximum primary current of 833.5 A and a consequent maximum secondary current of 16.67 kA.

After the predefined short-circuit duration of 60 ms, or six half-waves at 50 Hz, both thyristor switches (switch A and switch B) are opened and the experimental sequence is ended.

During the current limitation experiment the total secondary current is measured with a Rogowski-coil (D9) and a corresponding Rogowski-coil integrator (D10). The current distribution between the superconducting tapes in the secondary winding is measured with twelve individual customized Rogowski-coils (D11) and Rogowski-coil integrators (D14), identical to the setup described Chapter 5.2.2.

The integrated signal of the Rogowski-coils is recorded during the measurement using a transient recorder (D7) as well as a universal data acquisition device (D15), synchronized to the data acquisition of the transient recorder.

The primary voltage of the transformer during the short-circuit limitation is measured in order to calculate the grid impedance. The precise grid impedance is of importance to refine the simulation results shown in combination with the measurements as a part of Chapter 6.4. Since the phase-to-phase voltage of the 20 kV grid cannot be directly measured via a voltage divider, the voltage of both phases is determined with respect

to ground potential using two customized high-voltage dividers (D19). The transformer primary voltage can then be obtained by subtracting U_{ph1} from U_{ph2}.
In order to calculate the temperature of the superconductor in the secondary winding during the current limitation, the value of the secondary voltage needs to be measured as well. This is done with a differential probe (D8), whereat the output of the probe is connected to the transient recorder.
The transient recorder serves during this measurement as the main measurement controller. It synchronizes and triggers the two thyristor switches to the power grid and it triggers the data acquisition for the currents in the secondary winding as well as for the primary and the secondary voltage.

6 Test and Measurements on a 1 MVA-Class Current Limiting Transformer

This chapter presents the measurement results and corresponding discussions for the experiments, conducted on the superconducting transformer demonstrator. The experimental procedures and test setups are described in the previous Chapter 5.
The performance of the multi-layer insulation inside the walls of the vacuum cryostat is presented and discussed in section 6.1.
Section 6.2 shows the measurement results for the fundamental transformer parameters. These include the measurement of the short-circuit impedance of the transformer demonstrator, the current-sharing between the parallel superconducting tapes of the secondary winding and the iron core magnetization.
The results of the transformer loss investigation are presented and discussed in section 6.3. The no-load losses and the cryostat performance of the transformer demonstrator are shown in section 6.3.1. The measured losses at different AC-load levels are presented in section 6.3.2 and a method for separating the superconducting AC-losses from the measured total loss values is given in section 6.3.3. Section 6.3.4 introduces a method for up-scaling the measured transformer losses and the calculated efficiency from the single-phase 577.4 kVA transformer demonstrator to a three phase 1.73 MVA transformer model.
The last part of Chapter 6 presents the measurement results of the short-circuit current limitation measurements performed on the transformer demonstrator. The measurement results for the effectiveness of the current limitation are discussed and compared to predicted results, acquired in Chapter 2.3.2.

6.1 Performance of Multi-Layer Insulation

The measurements of the temperature at selected points in the multi-layer insulation inside the transformer cryostat walls are conducted during an alteration between no-load and nominal-load measurements on the transformer demonstrator. Figure 5.1 in Chapter 5.1 shows the sensor positions relative to the expected magnetic flux density during transformer operation.

The temperatures measured for the eight sensors are shown in Figure 6.1. The temperature axes (y-axes) for all sensors are plotted with the same scaling to make the temperature changes in the multi-layer insulation directly comparable between the different sensor positions.

For the first 100 minutes of the temperature measurements, the transformer is operated without current, neither in the primary nor in the secondary winding and therefore without a magnetic field, penetrating the multi-layer insulation. It can be observed that the temperature for all sensors during this period is decreasing. This is due to the fact that the transformer cryostat was manually refilled with liquid nitrogen 30 minutes prior to the temperature recording shown in Figure 6.1 and the cryostat assembly slowly adjusts to the change in the LN_2-level.

At 100 minutes after the start of the temperature recording, the transformer is electrically loaded by short-circuiting the secondary winding of the transformer and supplying the primary winding with the short-circuit voltage U_k. This causes nominal operating currents in both transformer windings to flow and consequently produces the full AC-magnetic field in the stray gap, as expected during nominal transformer operation. The period of nominal magnetic AC-fields in the transformer is indicated by a yellow background color in the graphs of Figure 6.1.

The temperature cannot be directly determined during the AC-loading of the transformer demonstrator due to interference between the measurement circuit and the magnetic stray field of the transformer. Thus, the AC-load phase is divided into 20 intervals of 5 minutes duration each. Between these intervals the AC-load is interrupted for 10 seconds to allow a precise temperature measurement. After the temperatures are acquired, the AC-load is continued. The short interruptions of the AC-load on the transformer are assumed to have no significant influence on the thermal behavior of the multi-layer insulation.

The AC-magnetic field causes eddy-currents in the multi-layer insulation. Depending on the orientation of the conductive layers of the multi-layer insulation foils, with respect to the magnetic field, the consequent Joule-heating effect varies. The highest temperature change during the measurement can be observed for the temperature sensor T5 with a increase by 5.9 K. Whereas no significant heating, cause by eddy-currents can be observed for sensor T7.

The slope of the temperature curves towards the end of the AC-load interval suggests that the temperature in the multi-layer insulation would further increase for AC-load intervals

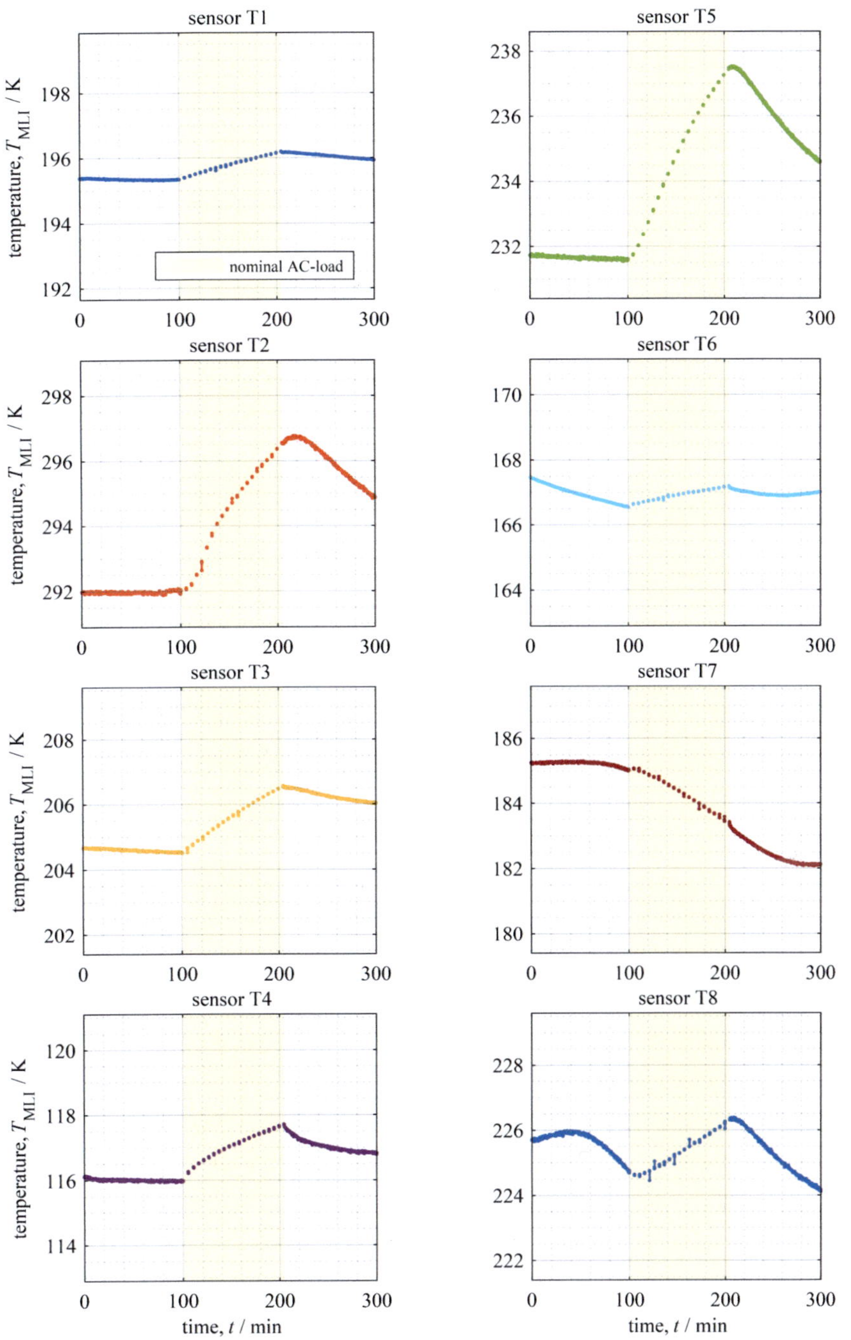

Figure 6.1: Temperatures at different positions in the multi-layer insulation, inside the vacuum gap of the GFRP-cryostat. The yellow background color indicates the operating phase during which the transformer is loaded with nominal AC-load, during the experiment time indicated by a white background color, the transformer is disconnected from the grid

with a longer duration.
The temperature increase in the multi-layer insulation can cause a reduction of the cryogenic insulation quality of the transformer cryostat during long-term operation. This issue needs to be addressed in further transformer design concepts, especially for a long-term grid operation.
One possibility for reducing the heating in the multi-layer insulation is the segmentation of the conductive layers. For this, the metallic coating of the single sheets of the multi-layer insulation are electrically separated into small segments using a CNC laser cutter. This segmentation does not affect the mechanical or thermal stability of the multi-layer insulation but it suppresses the formation of large loops of induced currents and therefore reduces the Joule-heating significantly.

6.2 Fundamental Transformer Parameters

6.2.1 Short-Circuit Impedance

The measurement method and electrical setup for the determination of the short-circuit impedance is discussed in Chapter 5.2.1.
The top graph of Figure 6.2 shows the full transient measurement of the primary transformer voltage and the current through the short-circuited secondary winding of the transformer. It can be seen that there is only a minor transient oscillation in the measured signals after the start of the measurement.
The lower graph in Figure 6.2 shows the last full half-waves for both, the primary voltage and secondary current. It can be seen that the measured primary transformer voltage is not purely sinusoidal but shows signs of higher frequency components. For this reason the short-circuit impedance is not calculated directly using the peak values of the primary voltage in the last half-wave but instead by using the RMS-values of the voltage signal, over the last ten acquired periods.
This yields a RMS-value for the primary transformer voltage of 577.8 V at a secondary short-circuit current of 577.7 A (RMS). The measured value for the secondary current is sufficiently close to the designed nominal current of 577.4 A, so that no further correction is needed and the relative short-circuit impedance can be directly calculated using Equation 5.2:

$$u_{\mathrm{k}} = \frac{577.8\,\mathrm{V} \cdot 100\,\%}{20\,\mathrm{kV}} = 2.89\,\%$$

The measured short-circuit impedance is lower than the short-circuit impedance, as it is

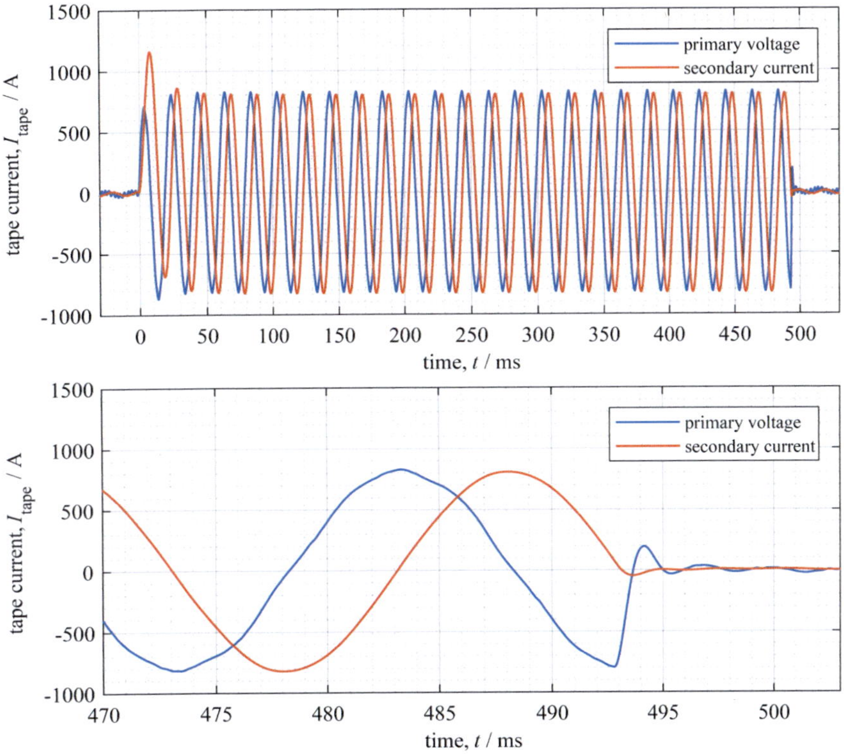

Figure 6.2: Primary voltage and secondary current with short-circuited secondary transformer winding. The measured values are used for the calculation of the short-circuit impedance u_k

calculated during the transformer design process (2.986 %), described in Chapter 2. The difference between the measured value and the calculation during the design process is 3.2 %.

6.2.2 Current-Sharing between Superconducting Tapes

The current-sharing between the parallel superconducting tapes in the secondary winding of the transformer demonstrator is shown in Figure 6.3. The top graph depicts the individually measured transient currents in the parallel superconducting tapes of the secondary winding while the transformer is connected to the 20 kV-grid and loaded with the nominal load resistor of 1.73 Ω.

The measured currents in the outer layer of tapes in the secondary winding, the tapes facing the stray gap of the transformer, are represented in the graph by continuous lines. The currents measured in the superconducting tapes facing the iron core, the inner layer of tapes, are represented by dashed lines respectively. The color of the lines in the graph

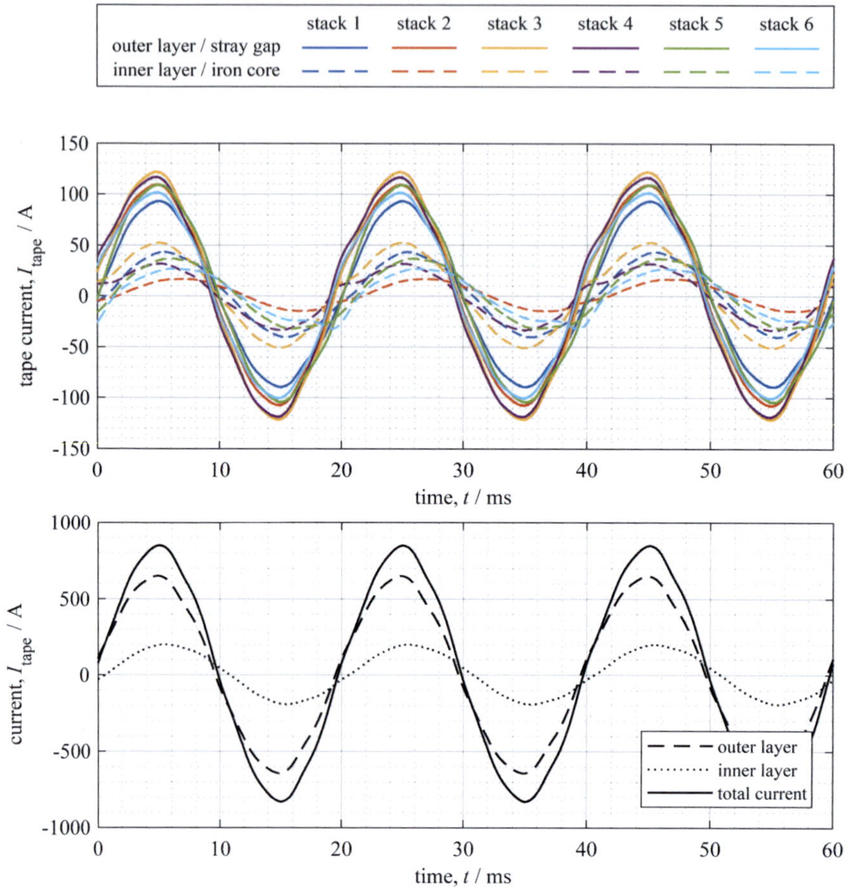

Figure 6.3: The top graph shows the current-sharing between the parallel tapes of the secondary winding during nominal operation of the transformer. The color of the lines in the top graph indicates the stack number, the tape belongs to. The bottom graph shows the sum of all currents in the outer layer, in the inner layer and in all conductors of the secondary winding

indicates which superconducting tapes and corresponding currents are arranged in the same stack of superconducting tapes.

It can be seen from the top graph of Figure 6.3 that the superconducting tapes in the outer layer of the secondary winding carry significantly larger currents in comparison to the tapes of the inner layer. This unequal current distribution can be observed as long as none of the single tape currents exceed the critical currents of the individual superconducting tapes. Because the electrical resistance of the parallel conducting paths in this operating condition is negligible, the current distribution primarily follows the induced voltages, as determined by the coil geometry. This means that a major share of the secondary current is induced in the outer layer of superconducting tapes in the secondary winding. This in turn causes a magnetic field which counteracts the primary magnetic field inside the secondary winding. The result is a partial compensation of the magnetic field in the outer layer of the secondary winding and, thus, in a lower induced voltage and a lower consequent current flow in the inner layer of superconducting tapes. The effect of unequal current distribution is less significant for normalconducting transformer windings since between normalconducting parallel wires the current distribution occurs mainly according to the resistance of the parallel conducting paths. For normalconducting parallel wires in a identical geometry, one would assume an almost homogeneous distribution of the secondary current between the outer and inner layer of conductors.

To avoid an unequal current distribution in conventional or superconducting transformer windings with parallel conducting paths, the conductors can be arranged in a continuously transposed Roebel-structure or in a otherwise transposed conductor arrangement [GFK+09, SGP+12].

The bottom graph in Figure 6.3 shows the summation of the currents in the outer and in the inner layer of superconducting tapes as well as the sum of all individually measured currents in the secondary winding.

In the measurement results, the sum of the currents in the outer layer of superconducting tapes is 452.6 A (RMS) whereas the inner layer of superconducting tapes carries a total current of 134.1 A (RMS). This corresponds to a current distribution between outer and inner layer of superconducting tapes of 77.15 % to 22.85 %.

The results further show that the currents in the secondary winding are not solely of sinusoidal nature. This is due to the higher frequency components in the grid voltage as it can be seen in Figure 6.2. The higher frequency components are observed as well during the measurements in the 20 kV-grid, shown in 6.4.

6.2.3 Iron Core Magnetization and No-Load Parameters

The measurement setup for determining the magnetization of the the iron core of the superconducting transformer demonstrator and the no-load parameters is described in Chapter 5.2.3.

The top graph in Figure 6.4 shows the measurement result for the magnetic flux density in the iron core of the superconducting transformer demonstrator during no-load operation at the nominal primary voltage of 20 kV (RMS).

The transient result for the magnetic flux density is acquired from the measured pick-up voltage $U_{\text{pick-up}}$ by the use of Equation 6.1.

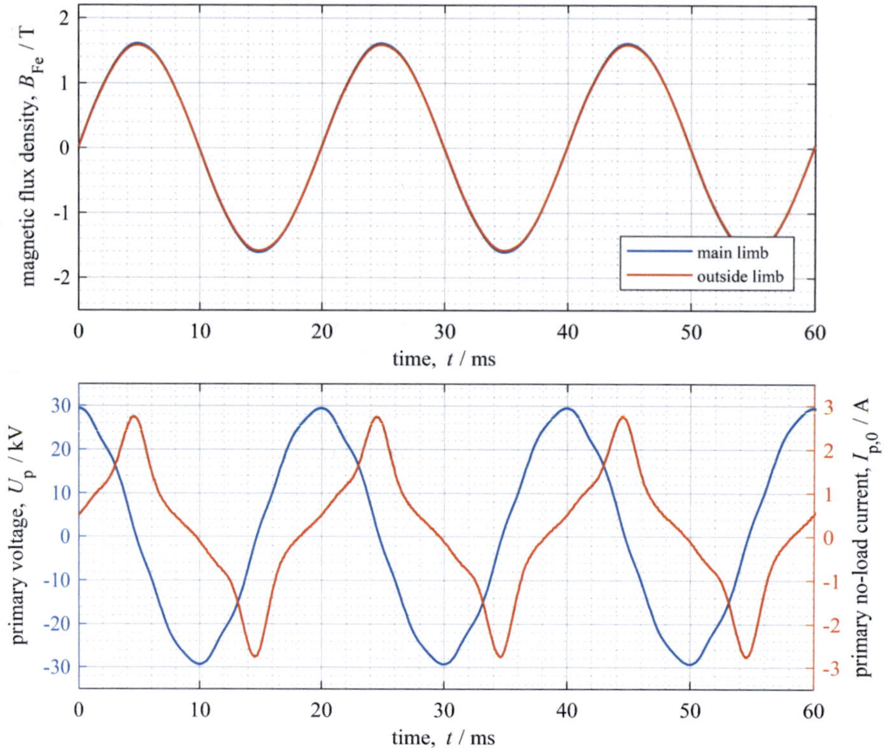

Figure 6.4: The top graph depicts the magnetic flux density in the iron core during nominal load operation of the transformer demonstrator. The bottom graph shows the corresponding primary voltage and primary no-load current

$$dB(t) = \frac{U_{pick-up}(t)\,dt}{n_{pick-up}\,A_{Fe,eff.}} \tag{6.1}$$

In this equation, dt represents the sampling time of the data acquisition device, $n_{pick-up}$ is the number of turns in the pick-up coils and $A_{Fe,eff.}$ is the effective cross section area of the iron core.

The acquired peak values for the magnetic flux density in the effective iron core cross section is 1.611 T for the main limb of the iron core, the limb carrying both the transformer windings. For the unwound limb a peak magnetic flux density of 1.592 T is measured. Both measured values for the magnetic flux density are in good agreement with the design value for the maximum magnetic flux density in the iron core, B_{max} of 1.6 T, as specified for the transformer design process, described in Chapter 2.3.1. The deviation from the design value amounts to 0.69 % for the main core limb and 0.50 % for the unwound core limb.

The bottom graph of Figure 6.4 shows the no-load operation parameters of the transformer demonstrator. It can be seen that the primary voltage during the measurement is not purely sinusoidal but shows higher frequency components together with the main frequency of 50 Hz. This is assumed to be an effect of the 20 kV power grid.
The pattern of the primary current is typical for the no-load operation of a transformer and it consists of the magnetization current of the iron core and of an additional capacitive current, reflecting the capacitance of the transformer primary winding [CFSW03]. The RMS-value of the measured no-load current is 1.375 A which equals 4.76 % of the RMS-value of the nominal operation current of 28.9 A. This ratio between no-load current and normal-operation current is relatively high for a transformer of the 1 MVA power-class [CFSW03] but can be explained with the high turn-to-turn voltage of the transformer design. The higher turn-to-turn voltage results in a increased iron core mass and therefore higher magnetization losses in the iron core.

6.3 Transformer Loss Investigation

This section presents the measurement results for the transformer losses. These include the losses in the primary transformer winding, the transformer iron core losses and the losses measured inside the transformer cryostat. An explanation on the measurements setup and methodology for the different transformer loss mechanisms is given in Chapter 5.3.
In the first part of this section the no-load performance of the transformer cryostat is presented and discussed. For those measurement results, two redundant calorimetric measurement techniques (loss-measurement A and loss-measurement B) are used, as described in Chapter 5.3.1.
Chapter 6.3.2 shows the results for the cryogenic losses inside the transformer cryostat, as they are acquired with the two measurement techniques for variable transformer loads, with and without the use of the additional calibration heater.
Section 6.3.3 explains the applied methodology for the separation of the total acquired cryogenic transformer losses into the different components, the AC-losses, the current lead losses, the resistive losses in the busbars of the secondary winding and the resistive losses in the pressure contacts of the secondary winding assembly.
The transformer loss investigation is concluded in Chapter 6.3.4 with the presentation of the applied approach to upscale the measurement results, acquired for the single-phase transformer demonstrator, to a three-phase unit.

6.3.1 Cryogenic Losses without Transformer Load

The losses inside the cryostat of the transformer demonstrator are determined by using the experimental setup as explained in Chapter 5.3.1. The experiment is conducted during no-load operation of the transformer, therefore the measurements only includes the two no-load loss components. These are the heat flow through the cryostat walls $\dot{Q}_{\mathrm{cryo}}$, also including the heat flow through the cryostat lid, and the heat flow along the brass current leads $\dot{Q}_{\mathrm{CL}}$.

The top graphs in Figure 6.5 shows a total of seven no-load measurements that were conducted as a part of the transformer loss investigation using loss-measurement A, as described in Chapter 5.3.1.

The measurements are represented by the different colored lines and show the no-load heat flow rate, $\dot{Q}_{\mathrm{in,0}}$ versus the level of liquid nitrogen inside the transformer cryostat. The liquid nitrogen filling level is the level of liquid nitrogen above the top end of the secondary winding assembly, given in millimeters. The presented measurements are not calibrated via the reference heater.

The results in the graph are acquired by calculating the heat input from the measured volumetric flow in the exhaust hose of the cryostat. This volumetric flow value can be converted into the actual heat flow rate using Equation 6.2.

$$\dot{Q}_{\mathrm{in,0}} = V_{\mathrm{N_2}}\, \rho_{\mathrm{N_2}}\, \Delta H_{\mathrm{vap,LN_2}} \tag{6.2}$$

In this equation, the volumetric flow rate acquired by the flow meter is represented by $V_{\mathrm{N_2}}$ and is given in liters of gaseous nitrogen per second. The specific density of gaseous nitrogen is given by $\rho_{\mathrm{N_2}}$ in kilogram per liter and $\Delta H_{\mathrm{vap,LN_2}}$ represents the enthalpy of vaporization of liquid nitrogen in Joule per kilogram ($\Delta H_{\mathrm{vap,LN_2}} = 197\ \mathrm{kJ/kg}$).

The results for the no-load heat flow rate $\dot{Q}_{\mathrm{in,0}}$ versus the liquid nitrogen level are in the range between 41 W and 106 W. The graph shows that the no-load heat flow rate is non-linear and strictly monotonic decreasing with the level of liquid nitrogen. This is an expected behavior since the decrease of the liquid nitrogen level inside the transformer cryostat allows the formation of thicker gaseous nitrogen layers, lowering the temperature gradient between the cryostat lid and the surface of the liquid nitrogen reservoir. This significantly reduces the heat flow into the cryostat through the cryostat lid and therefore improves the overall insulation quality of the transformer cryostat. The reduction of the liquid nitrogen level also reduces the temperature gradients inside the cryostat walls.

The average loss value versus the liquid nitrogen filling level is calculated from the single curves of the seven experiments. The average value is then fitted with a fifth-degree

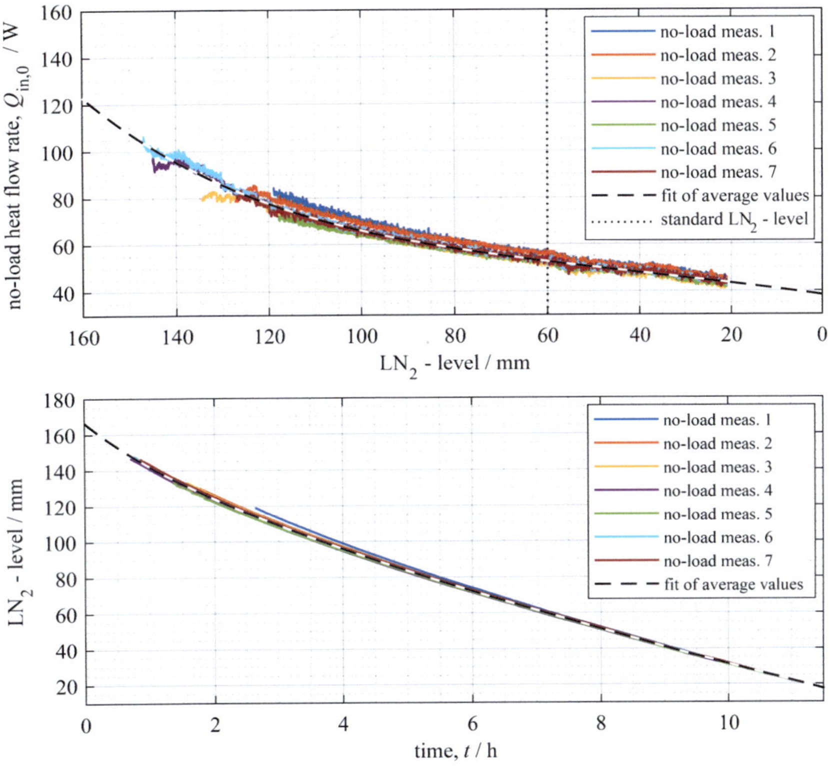

Figure 6.5: The top graph shows the measured no-load heat flow through the cryostat walls and the heat flow along the current leads into the cryogenic environment. The losses are calculated from the volumetric flow of gaseous N_2 leaving the cryostat and are not calibrated via the reference heater. The top graph further indicates the standard LN_2-filling level assumed for the loss calculations in Chapter 6.3.3. The bottom graph shown the corresponding LN_2-level versus the experiment time

polynomial fit. The fitted curve is shown as a dashed black line in the top graph of Figure 6.5. Represented by a dotted black line in the same graph, is the standard LN_2-filling level of 60 mm, as it is assumed for the calculation of the no-load cryogenic losses. These are presented as a part of the load dependent transformer losses in Chapter 6.3.3.
The bottom graph of Figure 6.5 shows the measurement of the liquid nitrogen level versus the experiment time as acquired by using loss-measurement B, described in Chapter 5.3.1. The colors of the lines in the graph correspond with the colors in the upper graph for the no-load heat flow rate.
It can be seen, that the change in the liquid nitrogen level is increased for higher LN_2-filling levels inside the transformer cryostat. This behavior is expected due to the higher no-load heat flow level for higher LN_2-filling levels.

For the later use of the liquid nitrogen filling level in the data analysis of the transformer loss measurements, the measurement points of the seven separately acquired no-load measurements are averaged and then fitted with a fifth-degree polynomial fit.

6.3.2 Cryogenic Losses at Variable Transformer Loads

The experimental setup used to measure the cryogenic losses at different transformer load levels is described in Chapter 5.3.1 and Chapter 5.3.2. During the measurement both measurement methods are used in parallel to acquire the total losses inside the transformer cryostat.
With the determination of the no-load heat flow rate for the volumetric flow measurement (loss-measurement A) and the no-load level change for the laser level measurement (loss-measurement B), described in Chapter 6.3.1, it is possible to separate the thermal conduction heat flow components ($\dot{Q}_{\text{cryo}}$ and $\dot{Q}_{\text{CL}}$) from the heat flow into the transformer cryostat as a consequence of current depending heating ($\dot{Q}_{\text{Cu}}$, $\dot{Q}_{\text{AC-loss}}$ and $\dot{Q}_{\text{contacts}}$).
For the measurement results presented in this chapter and in the following Chapter 6.3.3, the no-load loss components are subtracted from the presented values. This means that the results presented in the following only show the load dependent transformer loss components. These are the AC-losses in the superconducting tapes and the Joule-heating losses.

The load dependent heat flow $\dot{Q}_{\text{in,LMA}}$ based on loss-measurement A, is acquired from the measured volumetric flow V_{N_2}, using Equation 6.3. This equation and its components are identical to Equation 6.2 used to calculate the no-load heat flow into the transformer cryostat in Chapter 6.3.1.

$$\dot{Q}_{\text{in,LMA}} = V_{\text{N}_2}\, \rho_{\text{N}_2}\, \Delta H_{\text{vap,LN}_2} \tag{6.3}$$

To calculate the load dependent heat flow value $\dot{Q}_{\text{in,LMB}}$ from the change in the liquid nitrogen filling level acquired with loss-measurement B, Equation 6.4 is used.

$$\dot{Q}_{\text{in,LMB}} = \Delta L_{\text{LN}_2}\, A_{\text{LN}_2}\, \rho_{\text{LN}_2}\, \Delta H_{\text{vap,LN}_2} \tag{6.4}$$

The equation includes the measured change in the liquid nitrogen filling level, ΔL_{LN_2} in meters per second, the liquid nitrogen surface area inside the cryostat, A_{LN_2} stated in square meters, the density of liquid nitrogen, ρ_{LN_2} in kilogram per cubic meter and the enthalpy of vaporization for liquid nitrogen at ambient pressure in Joule per kilogram, represented by $\Delta H_{\text{vap,LN}_2}$.

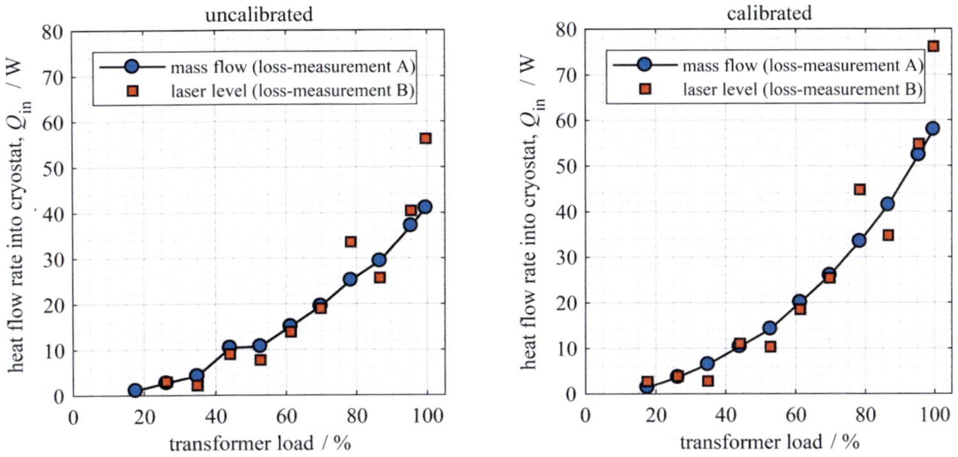

Figure 6.6: Measured load dependent cryogenic losses after removing the no-load heat flow through the cryostat walls and the heat flow along the current leads. The left graph shows the uncalibrated heat flow into the cryostat, whereas the measured values in the right graphs are values, calibrated via the the specified energy input from a reference heater

The left graph of Figure 6.6 shows the uncalibrated load dependent losses, calculated from the measured values. The blue circular markers represent measurement points at different transformer loads, acquired using loss-measurement A whereas the red squares represent measurement point acquired with loss-measurement B.
The graph on the right side of Figure 6.6 shows the load dependent losses inside the transformer cryostat after calibration via the calibration heater. For the calibration a reference heat flow into the cryogenic environment is provided by a defined value for the calibration heater power. This reference heat flow is used to calibrate the heat flow values, measured during the AC-load of the transformer.
The comparison between the uncalibrated and calibrated measurement values show a significant difference. In average the uncalibrated measurement values amount to 74.9 % of the calibrated values. The reason for the deviation between the measured and the reference values is explained in the following.

Systematic Deviation of Measurement Values

Figure 6.7 shows the thermodynamic T-H-diagram for nitrogen [JS73] at an ambient pressure of 1013.25 mbar. The enthalpy of vaporization $\Delta H_{\text{vap,LN}_2}$ is the enthalpy difference between the liquid saturation line and the vapor saturation line for a specific pressure. For the pressure of 1013.25 mbar the enthalpy of vaporization is marked in the diagram and has a value of $\Delta H_{\text{vap,LN}_2} = 197$ kJ / kg.

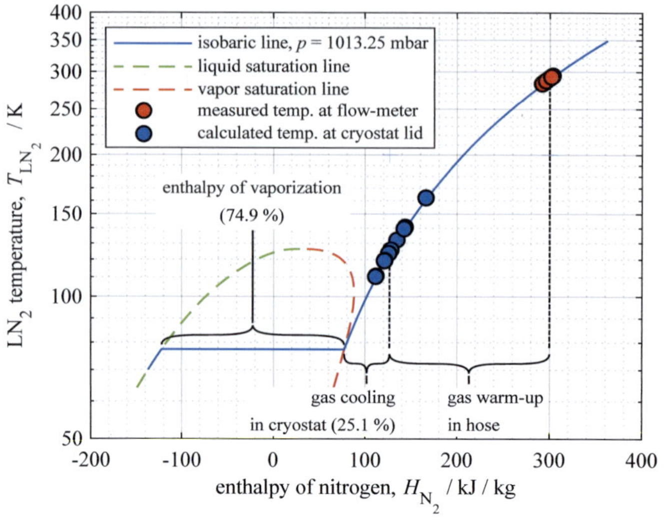

Figure 6.7: The graph shows the temperature-enthalpy diagram (T-H-diagram) for nitrogen. The blue line represents the isobaric line at a pressure of $p = 1013.25$ mbar. The enthalpy of vaporization $\Delta H_{vap,LN_2}$ is 197 kJ / kg. The temperatures measured during the calibration measurement at the flow meter and the calculated temperatures for the nitrogen leaving the cryostat are indicated by red and blue circular markers respectively

The results presented in the left graph of Figure 6.6 are derived from the measurement values, following Equation 6.3 and Equation 6.4 and by assuming an enthalpy of vaporization for the liquid nitrogen of 197 kJ / kg.

To assume a enthalpy of vaporization of 197 kJ / kg means that only the heat flow, dedicated for the evaporation of the measured amount of liquid nitrogen is taken into account. This further implies, that the gaseous nitrogen leaves the cryogenic environment at a temperature of 77.3 K, the temperature at the liquid saturation line in Figure 6.7. This assumption is certainly not correct, since the evaporated nitrogen inside the transformer cryostat continues to act as a gaseous coolant. This takes place for example due to the cooling of the current leads above the liquid nitrogen level, due to the cooling of the inside cryostat walls above the liquid nitrogen level and due to cooling of the inside-surface of the cryostat lid. The difference between the uncalibrated measured values and the calibrated values shows that an average value of 25.1 % of the total cooling inside the transformer cryostat is done by the evaporated gaseous nitrogen.

Figure 6.7 further shows the average measured temperatures at the flow meter during the calibration via the calibration heater as red circular markers, whereas blue circular markers indicate the calculated temperature of the gaseous nitrogen leaving the transformer cryostat. With the average calculated enthalpy of the nitrogen leaving the cryostat, the

unknown interval between the vapor saturation line and the enthalpy at room-temperature can be divided into two parts. The left part accounts for the enthalpy change due to gas cooling inside the cryostat and the right part represents the warm-up of the nitrogen gas inside the hose, connecting the flow meter to the cryostat lid.
For improving the reliability of the loss measurement, or, in order to not rely on an additional calibration heater, a precise measurement of the temperature of the evaporated nitrogen, leaving the transformer cryostat, would be required. With the additional determination of the exact temperature of the evaporated coolant, leaving the cryogenic system, it would be possible to directly calculate the heat flow into the system, without the need of any calibration method.
After the calibration of the measurements, the heat flow values acquired with loss-measurement A indicate a smooth, non-linear dependency on the transformer load. This is the expected behavior, since the Joule-heating losses scale quadratic with the current in the conducting materials and the AC-losses in the superconductor are dependent on the transport current via an exponent with a value of three or higher [Nor70, GZP^{+}14].
The majority of the values acquired with loss-measurement B are in good agreement with the results from loss-measurement A, with the exception of five measurement points. It is assumed that the reason for this disagreement are transient thermal contractions in the whole transformer demonstrator setup that occur during the AC-load measurements. This can be caused by not allowing the cryogenic system the time required to adjust to refilling liquid nitrogen, before the start of a measurement.

6.3.3 Separating Loss Components and AC-Losses

After the calibration of the measurement results, as shown in Chapter 6.3.2, the total measured heat flow can be separated into its single components, representing the different loss mechanisms inside the transformer cryostat. From this point on, the heat flow into the transformer cryostat will be referred to as transformer cryogenic losses.
The four main loss mechanisms inside the transformer cryostat are the:

- AC-losses in the superconducting tapes itself
- Joule-heating losses in the conducting material of the current leads
- Joule-heating losses in the copper busbars inside the secondary winding
- Joule-heating losses in the clamping contacts, connecting the superconducting tapes to the copper busbars

The AC-losses are the only loss-component that cannot easily be determined analytically. Whereas the three Joule-heating components can be calculated from the measured current at the different transformer load levels in combination with the predetermined electrical resistances.
The resistance value for the current leads is complex to determine, because it strongly depends on the liquid nitrogen filling level. For the separation of the AC-loss component presented in the following, the resistance values for the current leads at different transformer loads are not measured during the experiments but obtained from the numerical optimization model that was used to design the current leads, as described in Chapter 4.5.1. The resistance values are numerically calculated for each measurement point separately, depending on the actual liquid nitrogen filling level in conjuncture with the secondary transformer current and the duration of the experiment.
The resistance of the copper busbars inside the secondary winding assembly is calculated according to the cross section areas and dimensions of the bus bars and the resistivity of copper at a temperature of 77.3 K.
The contact resistance between the copper busbars and the superconducting tapes is measured during commissioning of the secondary winding, as a part of the critical current measurement of the single parallel conductors in the secondary winding assembly. These measured resistance values are directly used to calculate the losses in the individual contacts, depending on the secondary transformer current.
For the separation into the four main loss mechanisms it can be assumed that the difference between the sum of the three Joule-heating based loss contributions and the total measured losses inside the transformer cryostat is accounted for by the AC-losses in the superconducting tapes.
Figure 6.8 shows the total measured losses inside the transformer cryostat, depending on the transformer load, as blue circular markers. The total losses are separated into the three Joule-heating based contributions; the current lead losses, the losses in the copper busbars and the losses in the contacts. This leaves the difference between the sum of the three Joule-heating contributions and the total measured losses. This difference can be assumed to represent the AC-losses in the superconducting tapes.
Alternatively, the AC-losses inside the transformer cryostat can be estimated with a different approach. This is based on the fact that the Joule-heating losses inside the cryostat have a quadratic dependence on the secondary winding current whereas the AC-loss component in the superconducting tape depends on the secondary current following a power-law with an exponent of three or higher [GZP+14].
To apply this methodology for separation the Joule-heating losses from the AC-losses,

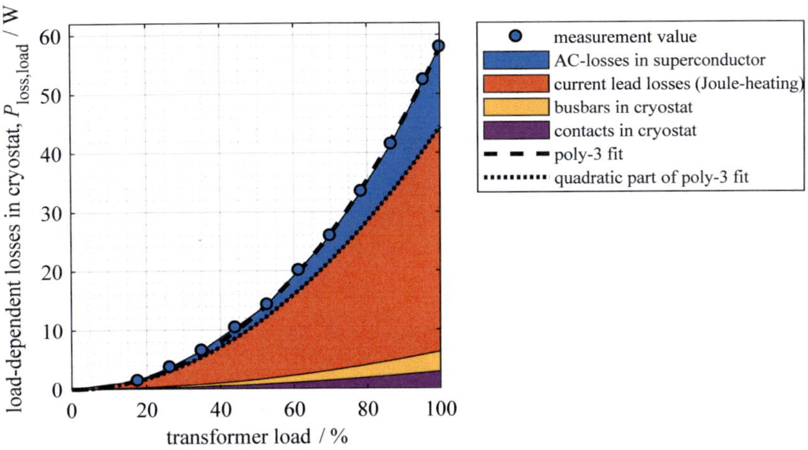

Figure 6.8: Load dependent loses $P_{\text{loss,load}}$ inside the cryostat of the transformer demonstrator. The total measured losses are separated into the different components in order to acquire the AC-loss contribution caused by the superconducting material in the secondary winding

the measurement values are fitted with a third degree polynomial fit function, as shown in Equation 6.5.

$$f(x) = p1\,x^3 + p2\,x^2 + p3\,x + p4 \tag{6.5}$$

To separate only between the quadratic and the cubic parts of the polynomial fit, the linear and the constant parameters, $p3$ and $p4$ are fixed to zero and therefore eliminated in Equation 6.5. The resulting fit has only the quadratic and cubic components with $p1 = 1.348e-05$ and $p2 = 0.004438$. The fit is presented in Figure 6.8 by a dashed black line, showing a good agreement with the measurement points.

The contribution to the total losses based on Joule-heating can now be separated by only plotting the polynomial function with the parameter $p2 = 0.004438$ and leaving the remaining parameters ($p1$, $p3$ and $p4$) set to zero. The result represents only the quadratic parts of the fitted function, the Joule-heating contribution, and is shown as a black dotted line in Figure 6.8.

The resulting separation of the AC-losses from the Joule-heating losses inside the transformer cryostat agrees well for both presented methods shown in Figure 6.8.

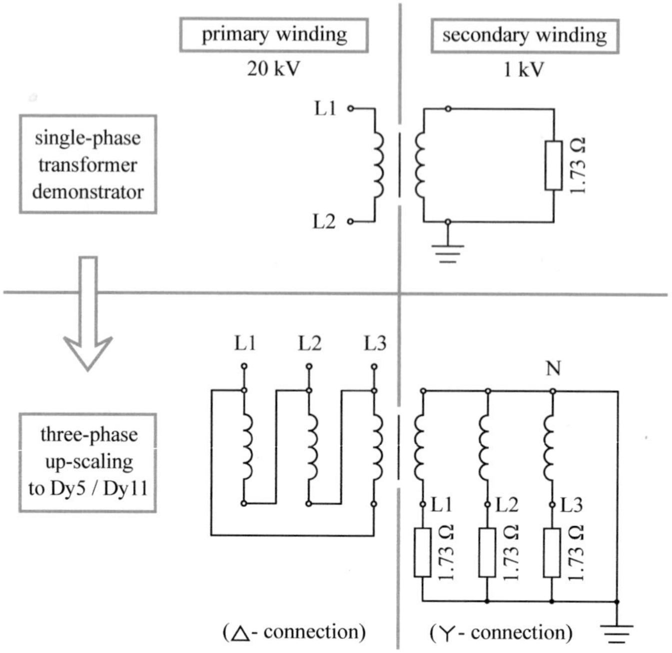

Figure 6.9: Single-phase circuit diagram of transformer test circuit (top) and three-phase up-scaling to Dy5 / Dy11-vector group (bottom). The Dy-vector group is typical for distribution transformers over a wide range of power-classes

6.3.4 Three-Phase Transformer Losses and Total Efficiency

The superconducting transformer demonstrator is designed as a single-phase unit, for the purpose of simplifying the transformer test circuitry and for reducing the overall complexity of the experimental setup. However, in order to examine and compare the total transformer losses, it is reasonable to up-scale the single-phase transformer demonstrator into a three-phase transformer model. This is done to allow a direct comparison of the transformer losses with conventional three-phase transformer designs.

A single-phase transformer is not ideal in terms of magnetization losses, since it requires a separate iron core limb only for the magnetic flux return. However, a separate limb for the magnetic flux return is not required in a three-phase design. This reduces the iron core mass to total transformer power ratio, and in turn, leads to a reduction of the losses for a three-phase transformer in comparison to a single-phase transformer design.

The two circuit diagrams, shown in the upper and lower part of Figure 6.9 present the single-phase connection scheme used for testing the transformer demonstrator and the three-phase connection scheme for the up-scaled model of the transformer losses.

The primary side of the single-phase transformer demonstrator is connected between two phases (L1 and L2) of the 20 kV-grid and the secondary side of the transformer is loaded with a nominal load resistor of 1.73 Ω. For a secondary winding voltage of 1 kV (RMS) this results in a complex nominal transformer power of 577.4 kVA (RMS).
The up-scaled three-phase model is connected on the primary side between the phases of the 20 kV-grid in a delta-configuration. The secondary transformer taps are connected in a wye-configuration to three separate 1.73 Ω load resistors. This results in a complex nominal transformer power of 1.732 MVA (RMS). The delta-wye-configuration is commonly used for distribution transformers of the 1 MVA-class.
For the up-scaling of the cryogenic system of the transformer, an arrangement of three individual cryostats for the superconducting windings of the three transformer phases is assumed. With this assumption the cryogenic losses and the conductive, as well as Joule-heating losses of the current leads triple for the three-phase transformer design.
The top graph of Figure 6.10 shows the total losses inside the three cryostats of the up-scaled three-phase transformer. The graph includes the four load dependent loss mechanisms as shown in Figure 6.8 together with the cryostat losses and the thermal conduction along the current leads. The latter two represent the no-load losses of the cryogenic environment of the three-phase transformer. At full load of the up-scaled three-phase transformer, the cryogenic losses account to 399.0 W whereas in no-load condition the cryogenic losses decrease to 225.3 W.
In the bottom graph of Figure 6.10 the total load dependent and no-load losses of the up-scaled three-phase transformer are shown. For this visualization the cryogenic losses are scaled with an efficiency of a cryogenic refrigeration system η_{cooling} of 6.5 %.
The graph includes the load dependent losses in the primary winding and the non-load-dependent losses in the iron core. The total losses for the up-scaled three-phase transformer at 100 % nominal load are determined to be 16.05 kW whereas the the losses are reduce to 10.58 kW during no-load operation of the up-scaled three-phase transformer.
During the up-scaling, the complex nominal transformer power triples from 577.4 kVA to 1,732 kVA while the transformer iron core mass only increases by the factor of 1.759 from 4,001.3 kg to 7,039.8 kg. This is due to the change of the transformer iron core design from a single-phase U-I-type, to a E-I-type for the up-scaled three-phase design. Since the iron core losses are the largest contributor to the total transformer losses, the up-scaling from a single-phase to a three-phase design increases the total transformer efficiency.
Table 6.1 shows the numeric loss values, corresponding to the bottom graph, presented in Figure 6.10.

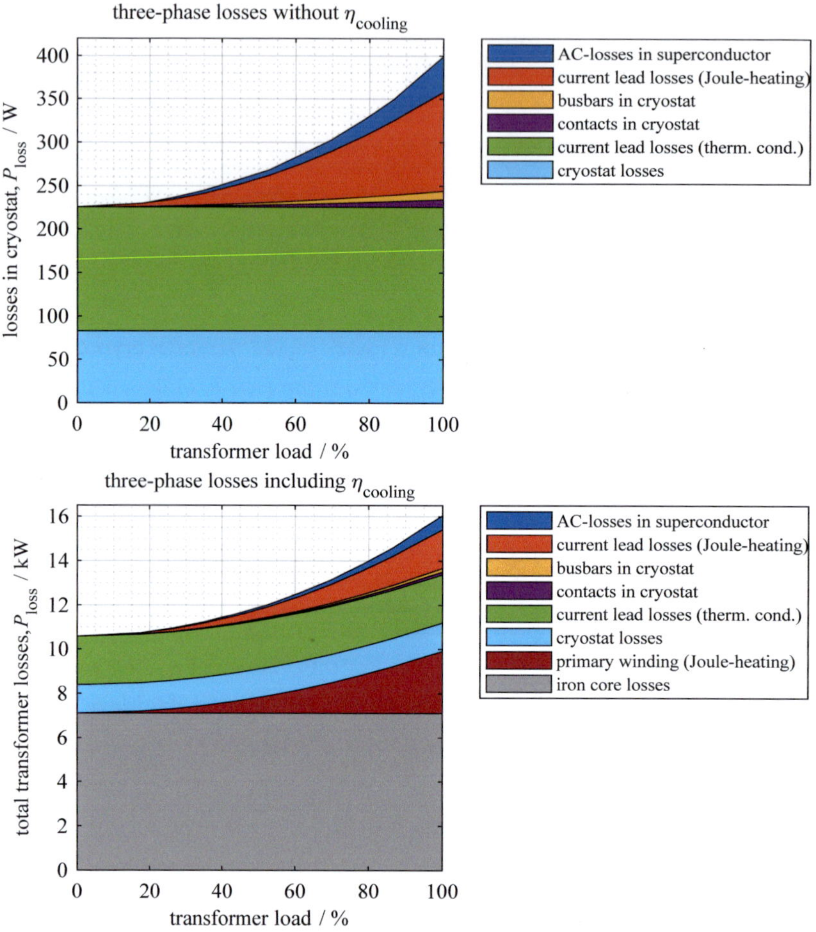

Figure 6.10: Losses $P_{loss,total}$ of the transformer demonstrator, up-scaled to a three-phase unit. The top graph depicts only the losses inside the three transformer cryostats, whereas the bottom graph shows the total transformer losses of a three-phase unit, including the efficiency of the cooling for the superconducting windings $\eta_{cooling} = 6.5$ %, the losses in the primary winding and the iron core losses

Table 6.1: List of numeric values for the total losses $P_{loss,total}$ of the transformer demonstrator, up-scaled to a three-phase unit

transformer load / %	0	17.6	26.3	35.0	44.0	52.7	61.3	69.8	78.4	86.7	95.5	100
secondary transformer current / A	0	101.9	152.1	202.0	254.3	304.2	354.1	403.3	452.9	500.4	551.2	577.4
AC-losses in superconductor * / W	0	2.74	30.12	50.86	84.72	92.26	157.9	201.0	286.1	374.5	547.9	633.3
current lead losses (Joule-heating) * / W	0	53.53	120.3	213.0	338.2	484.8	657.7	854.3	1078	1319	1603	1750
busbars in cryostat * / W	0	4.86	10.83	19.10	30.26	43.31	58.67	76.10	95.95	117.1	142.1	155.0
contacts in cryostat * / W	0	4.19	9.33	16.47	26.09	37.34	50.58	65.61	82.72	101.0	122.5	133.6
current lead losses (therm. cond.) * / W	2188	2188	2188	2188	2188	2188	2188	2188	2188	2188	2188	2188
cryostat losses * / W	1277	1277	1277	1277	1277	1277	1277	1277	1277	1277	1277	1277
primary winding (resistive) / W	0	87.66	195.4	344.7	546.0	781.5	1058	1373	1731	2114	2565	2797
iron core losses / W	7110	7110	7110	7110	7110	7110	7110	7110	7110	7110	7110	7110
total transformer losses / W	10577	10729	10942	11220	11601	12015	12560	13146	13851	14602	15558	16045
resulting three-phase efficiency / %	-	96.49	97.60	98.15	98.48	98.69	98.82	98.92	98.98	99.03	99.06	99.08

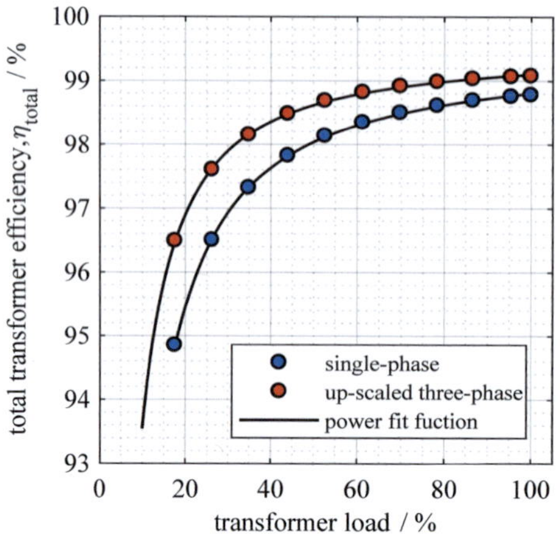

Figure 6.11: Total efficiency η_{total} of the up-scaled three-phase transformer versus the transformer load

Figure 6.11 shows the load dependent efficiency of the single-phase transformer demonstrator represented by blue circular markers. It can be seen that an efficiency of 98.78 % is reached if the single-phase transformer is operated at 100 % nominal load whereas the transformer efficiency sharply decreases for lower transformer load values. This is an expected behavior for superconducting as well as for conventional transformers which is caused by the non-load-dependent losses of the iron core.
The graph in Figure 6.11 further shows the efficiency of the up-scaled three-phase transformer as red circular markers. The maximum efficiency of the three-phase transformer of 99.08 % is reached if the transformer is loaded with 100 % of the nominal load. The efficiency for load values below 100 % of the nominal transformer load generally follows the trend of the efficiency of the single-phase unit but still shows higher efficiency values for lower load values. The increasing discrepancy between the efficiency of the single-phase design and the efficiency of the three-phase design for lower transformer load values is caused by the changed ration between the iron core mass and the transformer nominal power.

6.4 Current Limitation

6.4.1 Measured and Simulated Current Limitation

The simulation of the current limitation behavior shown in this chapter, follows the methodology presented in Chapter 2.2.2. However, the simulation parameters for the average critical current density, $J_{c,aver.}$ and for the temperature dependent tape resistance,

Table 6.2: List of updated parameters to refine the simulation results for the current limitation behavior

parameter description	symbol	original value	updated value
primary voltage	U_p	20 kV	20.655 kV
average critical current density	$J_\text{c,aver.}$	140 A	160.58 A
copper stabilization thickness	th_Cu	110 µm	105.25 µm
substrate layer thickness	th_HY	50 µm	48.0 µm
temperature dependent tape resistance	$R_\text{tape}(T)$	from material properties	measurement for tape 1 in Figure 3.4

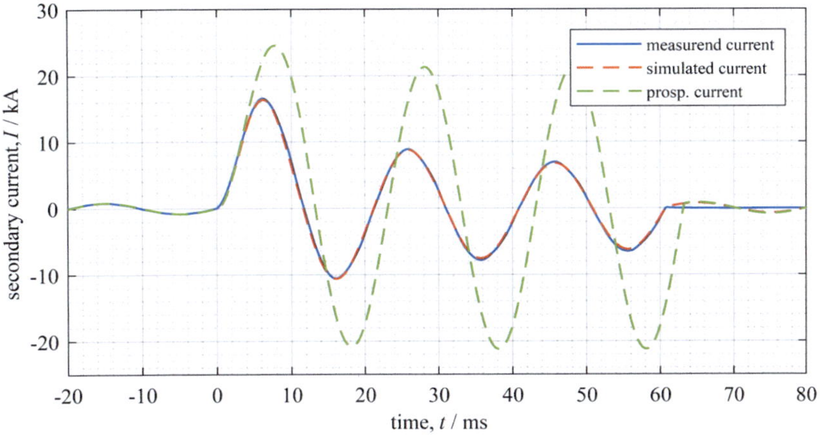

Figure 6.12: Comparison between measured and simulated secondary current in the transformer winding I_s. To show the effectiveness of the current limitation, the simulated prospective current is additionally visualized

$R_\text{tape}(T)$ are updated according to the preliminary measurement results as described in Chapter 3. The values for the thickness of the copper stabilization and the substrate layer are updated based on conducted microscopic examinations on the tapes sectional area. The value for the primary transformer voltage U_p is updated by the value for the grid voltage, as measured during the actual current limitation measurement on the transformer demonstrator. The updated values for the simulation parameters are listed in Table 6.2.

Figure 6.12 shows the measurement result for the current limitation characteristics of the transformer demonstrator, obtained with the measurement setup as it is described in Chapter 5.4. During the current limitation measurement the transformer is loaded over a duration of 60 ms with the short-circuit resistor with a resistance value of 25.3 mΩ.

In addition to the measured secondary current, Figure 6.12 also shows the simulated secondary current. The simulation of the secondary current is based on the updated simula-

Table 6.3: Absolute peak measurement values, prospective current values and simulated values for each of the the six half-waves of the current limitation experiment. The values shown in the columns marked with [1] are generated with the updated simulation parameters presented in Table 6.2 whereas the values in the columns marked with [2] are based on the original design parameters introduced in Chapter 2.3

half-wave	measured secondary current I_s	simulated prospec-tive current	limitation compared to prospective current	simulated secondary current[1]	derivation from measure-ment[1]	simulated secondary current[2]	derivation from measure-ment[2]
1st	16.58 kA	24.52 kA	67.62 %	16.38 kA	-1.22 %	16.67 kA	0.54 %
2nd	10.61 kA	20.84 kA	50.91 %	10.65 kA	0.38 %	10.73 kA	1.12 %
3rd	8.86 kA	21.27 kA	41.65 %	8.75 kA	-1.26 %	8.94 kA	0.89 %
4th	7.86 kA	21.22 kA	37.04 %	7.58 kA	-3.69 %	7.77 kA	-1.16 %
5th	6.94 kA	21.23 kA	32.69 %	6.78 kA	-2.36 %	6.96 kA	0.29 %
6th	6.18 kA	21.23 kA	29.11 %	6.54 kA	5.50 %	6.35 kA	2.68 %

tion parameters shown in Table 6.2. The figure further shows the simulated prospective current for a conventional transformer with identical specifications.

To allow a better comparison, Table 6.3 lists the absolute peak values for each half-wave in the measured secondary current, the simulated secondary current and for the the simulated prospective current, as they are shown in Figure 6.12.

The listed values show, that the exhibited current limitation of the transformer demonstrator reduces the secondary current in the peak of the first half-wave to 16.58 kA. This accounts for a current limitation back to 67.62 % of the simulated prospective peak current of 24.52 kA. The current limitation increases during the course of the experiment, until the limited current is reduced to a value of 6.18 kA in the sixth half-wave, what equals 29.11 % of the predicted prospective current of 21.23 kA.

As it can be seen from the values in row five and row six of Table 6.3, the simulation with the updated parameters matches the measurement results well and the simulated peak values for the secondary current show a maximum deviation from the measured values between -3.69 % and 5.50 %. The table further presents, in column seven and column eight, the simulation results acquired during the transformer design phase, described in Chapter 2.3.3. These results are based on the original design parameters and show a deviation from the measured secondary current between -1.16 % and 2.68 %.

This comparison suggests that an adjustment of the simulation parameters to the measured operational values or to the exact material properties is not necessarily required for the presented case. The simulation yields useful results even if it is solely based on the given transformer design parameters and on literature values for the temperature dependent tape resistance.

6.4.2 Temperature Increase in Secondary Winding

During the measurement of the current limitation of the transformer demonstrator the primary voltage is recorded in addition to the secondary transformer current. With the value for the measured primary voltage it is possible to determine the expected voltage drop on the secondary transformer taps during the current limitation. In combination with the known resistance values for the current leads, the copper busbars and the contacts in the secondary winding arrangement, the voltage drop on the secondary transformer taps can be used to extract the resistance value for the superconducting tapes in the secondary winding.

With the temperature dependent superconducting tape resistance $R_{\text{tape}}(T)$, presented in in Figure 3.4 of Chapter 3.2, the determined resistance value for the superconducting tapes in the secondary winding can be converted into the temperature value of the superconducting tapes, T_{tape}. However, this only works under the assumption of a homogeneous temperature distribution along the superconducting tapes.

The top graph of Figure 6.13 shows the temperature of the superconducting tapes determined from the measurement values at the peak of every half-wave and the simulated temperature increase during the current limitation. The simulation of the temperature increase is based on the updated simulation parameters presented in Table 6.2.

The figure additionally shows the maximum predicted superconductor temperature as it is simulated during the transformer design phases, presented in Chapter 2.3. The maximum predicted superconductor temperature is based on the original design parameters and therefore deviates from the values based on the adjusted parameters.

It can be seen from the graph that the measured temperature agrees well with the simulation results, for the first three half-waves. The measured temperature of the superconducting tape for the fourth, fifth and sixth half-waves are below the simulated temperature values. The measured temperature in the peak of the sixth half-wave is 307.3 K whereas the simulated temperature for the sixth half-wave is 326.8 K.

The significant difference between measurement value and the simulation value of 19.5 K is assumed to be caused by the absence of the thermal conductivity properties of the winding former in the simulation model. For the simulation the only cooling applied to the system of the superconducting tape is the heat transfer to the surrounding liquid nitrogen. However, the real tape during the measurement is also in direct contact with the GFRP-winding former what allows additional cooling of the superconducting tape by a

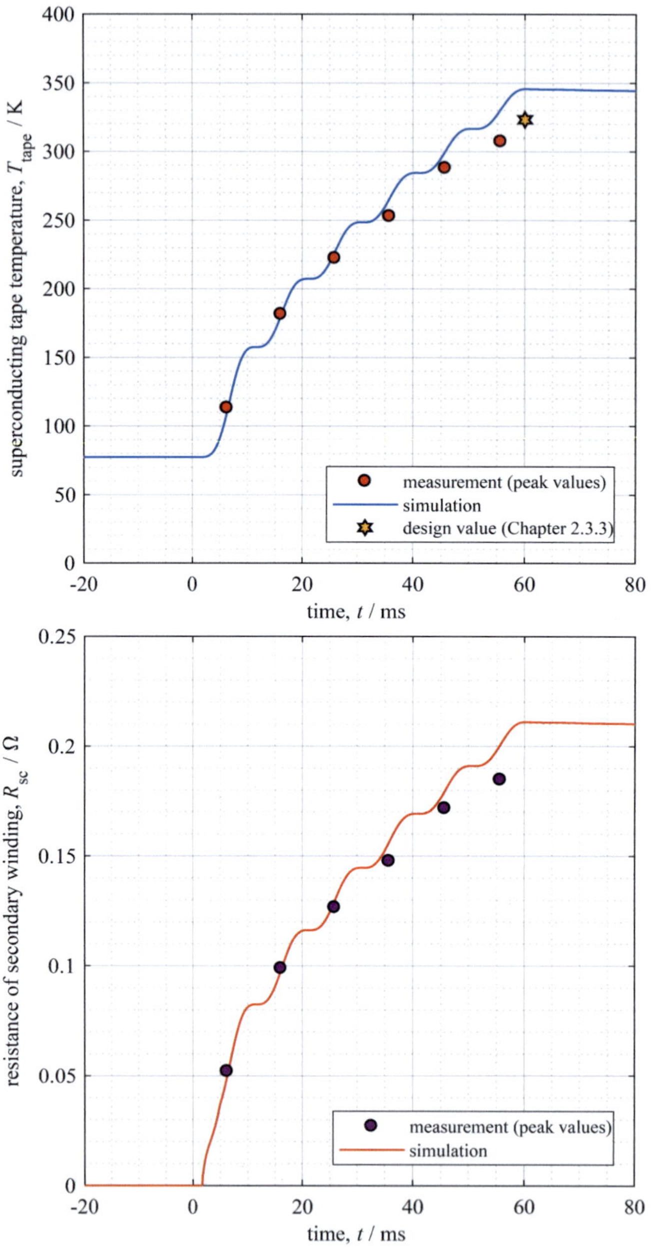

Figure 6.13: The top graph shows the measured (red circular markers) and simulated (blue line) temperature of the superconducting tape during the fault current limitation. The temperature points are calculated from the peak values of the measured secondary current and primary voltage during the experiment. The yellow marker represents the maximum simulated tape temperature acquired during the transformer design phase. The bottom graph shows the increase in the resistance of the the secondary winding R_{sc}. The measured values are represented by purple markers and the simulated resistance value is plotted as a red line

direct heat flow into the material of the winding former.

The bottom graph of Figure 6.13 shows the measured increase in the resistance of the superconducting part of the secondary transformer winding. The depicted values are determined from the measurement values at the peak of every half-wave and show the total resistance values for the twelve superconducting tapes connected in parallel. The graph also shows the simulated increase in the resistance.

The increasing disagreement in the resistance, between the measured and simulated values, towards the end of the current limitation is assumed to be also caused by the imperfect representation of the cooling inside the secondary winding arrangement in the numerical simulation model.

6.4.3 Current-Sharing during Current Limitation

During the current limitation measurement the secondary current is acquired individually for each of the twelve parallel superconducting tapes of the secondary winding. The twelve individual tape currents during the current limitation are presented in the top graph of Figure 6.14. The currents represented by continuous lines are currents measured for tapes in the outer layer of the secondary winding whereas the currents represented by dashed lines are measured for the inner layer of superconducting tapes. The colors of the lines in the graph indicate the six different stacks of superconducting tapes.

It can be seen that, during the short-circuit experiment, the current distribution between the single superconducting tapes is inhomogeneous. The current ranges from 1029 A to 1718 A in the peak of the first half-wave and from 402.7 A to 684.4 A in the peak of the sixth half-wave. The ratio of the current distribution between the parallel tapes remains almost identical during the course of the current limitation.

The inhomogeneous current distribution during normal operation is caused by the dominant unequal contact resistance and the resistance of other normal conducting components in the parallel current paths, as described in Chapter 6.2.2. However, the situation is different after the quench of the superconducting tape during the current limitation. Here, the resistance of the quenched tapes becomes dominant and it is, with the values shown in Figure 6.13, about three magnitudes higher than the average resistance of the contacts (106.4 $\mu\Omega$) and normal conducting components (121.6 $\mu\Omega$) in the current paths. For this reason it is assumed that the unequal current distribution during the current limitation is caused by unequal temperatures of the parallel superconducting tapes and consequently differences in the resistance of the superconducting tapes.

The graph on the bottom left of Figure 6.14 shows a magnification of the current distribution between the parallel superconducting tapes in the first half-wave of the current

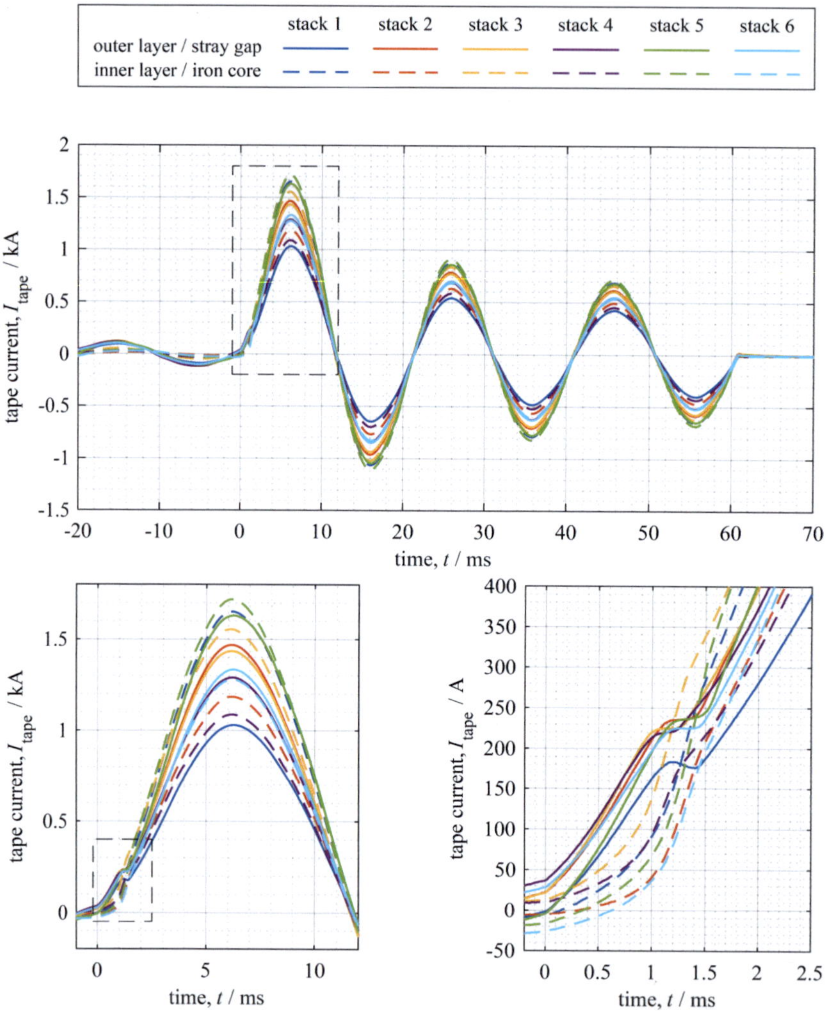

Figure 6.14: Current-sharing between parallel superconducting tapes in the secondary transformer winding. In the first 1.2 ms after the star of the short-circuit, more current is carried by the superconducting tapes in the outer layer. After this point all tapes in the winding are carrying currents larger than their respective critical current I_c

limitation. The graph on the bottom right of Figure 6.14 presents a further magnified view, depicting only the first 2.5 ms after the start of the short-circuit. From this graph it can be seen, that the current distribution in the first millisecond after the start of the short-circuit follows the same pattern as described in Chapter 6.2.2. The current in the secondary winding is mainly transported in the outer layer of superconducting tapes and the inner layer experiences a lower induced voltage and therefore lower transport currents.

This pattern changes between 1 ms and 1.5 ms after the start of the short-circuit limitation. During this period of the experiment the currents in the single tapes of the outer layer exceed their individual critical currents and therefore the superconducting material starts exhibiting an electrical resistance. This leads to a commutation of the current into the, at this point still fully superconducting, inner layer of tapes in the winding and to a consequent stagnation of the current in the outer layer. This commutation process is concluded about 1.5 ms after the start of the short-circuit. From this point on, the current distribution is assumed to solely follow the different resistance values of the parallel current paths and the resistance values exhibited by the superconducting tapes themselves.

6.4.4 Current Limitation for Different Fault Conditions

The results of the measurements on the current limitation presented in this chapter are acquired with a short-circuit resistor of 25.3 mΩ, as described in Chapter 5.4. With the use of the simulation model for the transformer demonstrator, utilizing the updated parameters shown in Table 6.2, it is possible to expand the conducted measurements towards other short-circuit resistances and also for a full short-circuit on the secondary transformer taps.

Figure 6.15 shows the results of the additional simulation for a wide range of short-circuit loads, between the resistance of the nominal load of 1.73 Ω and a full short-circuit load of 0 Ω.

The top graph presents the simulation of the limited transformer current versus the prospective current. The different colored lines represent the absolute peak current values for a specific half-wave of the current limitation whereas the prospective current is represented by a dashed black line. The ends of the lines on the right hand side of the graph indicate the simulated current values for a short-circuit resistance of 0 Ω. It can be seen that a significant current limitation takes place only for short-circuit currents exceeding 5 kA. The measurements are marked in the graph by colored circular markers, whereas the nominal operation point of the transformer is indicated by a black, diamond shaped marker.

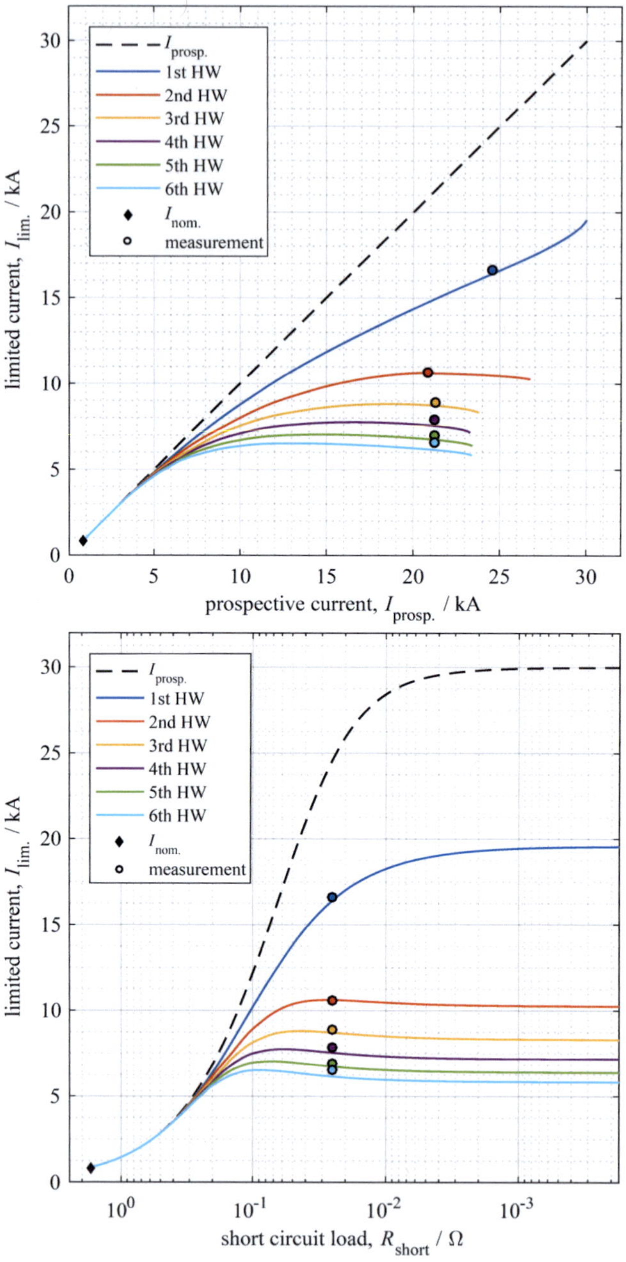

Figure 6.15: The top graph shows the simulated limitation of fault currents in the different half-waves of the fault together with the prospective current. The expected currents are plotted versus the prospective current. The circular markers visualize the measured current limitation during a short-circuit via a resistor of 25.3 mΩ. The bottom graph presents the simulated limited fault current versus the resistive load value causing the fault

The bottom graph of Figure 6.15 shows the simulation of the limited transformer current versus the short-circuit resistance. The colored lines represent the absolute peak current values for the different half-waves of the current limitation whereas the prospective current is represented by a dashed black line. The measurement values for the limited short-circuit current in the different half-waves are represented in the graph by colored circular markers and the nominal operation point is represented by a black diamond shaped marker.

The bottom graph of Figure 6.15 allows the direct prediction of the short-circuit current and the resulting effectiveness of the current limitation of the transformer demonstrator depending on the value for the applied short-circuit resistance.

7 Summary, Conclusion and Outlook

Superconducting transformers with fault current limiting properties can play an important role in managing the challenges that have to be faced when designing the power grids of the next generations or for improving and modernizing existing grid topologies. They can enable grid interconnections that simply were not possible before, or were economically not viable with the available conventional technologies for fault current control and protection.

Superconducting current limiting transformers provide an attractive combination between two technologies - the superconducting transformer and the resistive superconducting fault current limiter. Research on superconducting transformers, with and without the capability of fault current limitation, has shown that it is easily possible to combine the two main benefits offered by the novel properties of modern superconductor material. These benefits are a significant reduction of the Joule-heating losses and the current-triggered resistance increase of the superconducting material. The result of this combination is a superconducting transformer that can be significantly lighter, more compact and can have a higher efficiency in comparison to a conventional transformer. It further offers a fail-safe, efficient and fast limitation for short-circuit currents in a power grid.

The results of the research that was done during the course of this thesis further support this statement and can be concluded as follows:

Firstly, a design process for superconducting fault current limiting transformers was developed. This design process includes approaches, originally established for conventional transformers. These approaches were modified to be used for superconducting transformers and to facilitate the accessibility of design features offered by the novel properties of modern superconducting materials. The developed design tools not only allow the dimensioning and optimization of superconducting current limiting transformers but also enable a detailed analysis of the current limiting behavior and the consequent grid interaction. This was achieved by implementing a transient simulation model into the otherwise static design process.

The transformer design tool was used to generate an optimized transformer design for demonstrating the feasibility of a fault current limiting superconducting transformer of the 1 MVA power-class.
Secondly, experiments were conducted to study the general behavior of thin film superconducting tapes in over current conditions, as they appear in a fault current limiter during a short-circuit event. As a second part of the preliminary experiments, measurements were carried out to evaluate the heat exchange process between a quenched superconductor and adjacent liquid nitrogen as a coolant. The results of these preliminary superconductor evaluations agree well with the simulations resulting from the transformer design process. The acquired results were further used to choose the best suited commercially available superconducting tape for the developed transformer design.
Thirdly, a single-phase superconducting demonstrator was constructed and manufactured, according to the design that was developed during the transformer design process. The transformer demonstrator consists of a conventional resin-cast 20 kV primary winding and iron core, both operated at room-temperature. The secondary winding of the transformer is a superconducting winding, designed for effective current limitation and operated at 1 kV. The superconducting winding is wound on a patented winding former, for protecting the superconductor material against magnetic forces during the fault current limitation. For cooling to the operation temperature of 77.3 K, the secondary transformer winding is immersed into an open bath cryostat, containing liquid nitrogen at ambient pressure.
Fourthly, measurements were conducted on the manufactured transformer demonstrator. These covered the following operational properties of the transformer during nominal operation as well as during a fault in the power grid:

- The performance of the developed transformer cryostat and the utilized thermal insulation with sheets of multi-layer insulation - The measurements results prove the expected thermal insulation properties of the designed vacuum insulated cryostat with a heat flux density through the cryostat walls of less than 20 W/m^2. The temperature measurements in the multi-layer insulation further demonstrate the necessity for preventing Joule-heating inside the insulation material itself.

- Measurements of the total transformer losses and a later separation into the different loss contributions in the setup of the transformer demonstrator - These measurements show a total transformer efficiency of 99.08 % for a three-phase transformer model at full load. They further indicate that the AC-losses account to 633 W for a three-phase model under full load.

- Evaluation of the normal operation parameters of the transformer - The measurements verify the expected values for the no-load losses. Also, the measured short-circuit impedance meets the design value of 2.986 % with a deviation of only 3.2 %.
- Fault current limitation experiments to determine the current limitation behavior and the quench of the superconducting transformer winding - The results of the measurements show a current limitation to 67.62 % of the prospective current in the first half-wave and to less than 29.11 % in the last half-wave of the fault. The measured current limitation agrees very well with the simulated results and shows a maximum deviation of 5.5 %.

All acquired results show a satisfying degree of agreement between the predicted transformer parameters and the measured values. This confirms the accuracy of the transient simulation and prediction model applied during the transformer design process. It also shows that the design tool is a capable instrument for developing superconducting transformers and that it provides a valid solution for the otherwise complex process of designing superconducting fault current limiting transformers.

The next step towards a commercial superconducting current limiting transformer for long-term grid operation would be a further up-scaling of the transformer demonstrator to a three-phase unit of the 10 to 20 MVA-class. This transformer should include a sophisticated refrigeration system with closed-loop cooling of the superconducting current limiting windings, designed for long-term operation. Such a refrigeration system would further increase the total transformer efficiency and would allow a demonstration in a real power grid. A successful grid demonstration over an extended time period would further prove the reliability, accessibility and attractive advantages of the developed technology for superconducting current limiting transformers.

8 Appendix

8.1 Optical High-Speed Measurement Setup

Figure 8.1 shows a cross section view of the measurement setup used for the acquisition of the high-speed video data presented in Chapter 3.3, Chapter 3.4 and Chapter 8.2.

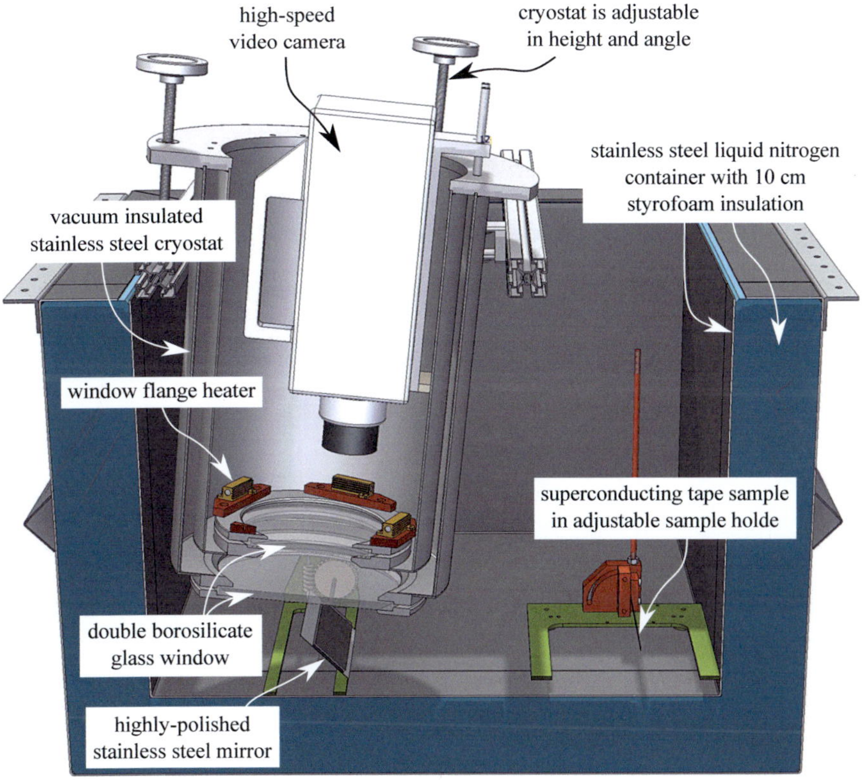

Figure 8.1: Measurement setup for acquiring high-speed video of quenching superconducting tapes in a open bath of liquid nitrogen at ambient pressure (cutaway view of a computer aided visualization)

8.2 1D-Recovery Simulation

For the numerical simulation of the quench propagation and recovery-under-load process, a 1-dimensional model of a superconducting tape is used. The single dimension represents the tape length, since the temperature distribution along the tape is inhomogeneous whereas the temperature over the cross section area of the tape is assumed to be negligible. The inhomogeneous temperature distribution along the tape is shown in the high-speed video recordings of the recovery process, in Chapter 3.3 and Chapter 3.4.
The simulation is conducted by solving the heat equation for a finite number of elements distributed along the tape length, with a forward-Euler method. This is implemented similar to the simulation for the transient current limitation, shown in Figure 2.6. However, the 1-dimensional tape model additionally requires the temperature dependent thermal conductivity $k(T)$ between the finite elements along the tape.
For the simulation the superconducting tape is quenched with a sinusoidal current of $I_{\text{lim,peak}} = 800$ A at 50 Hz for a duration of 100 ms. After the quench, the tape recovers under load while carrying a recovery current of $I_{\text{rec,peak}} = 197$ A.
Figure 8.2, Figure 8.3 and Figure 8.4 show a comparison between the optically acquired bubble coverage array and the simulated temperature distribution over time, versus the tape position. The comparison shows that the simulation of the temperature distribution over time and tape position agrees well with the corresponding bubble coverage arrays.
The maximum simulated temperature reached before the full recovery of the tape is 123.6 K for the tape with two solder dots, 112.0 K for the tape with four solder dots and

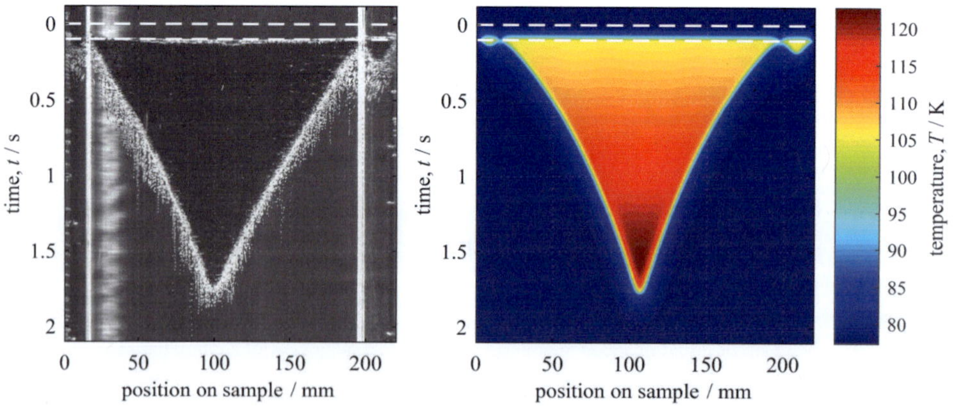

Figure 8.2: Bubble coverage array of a recovering superconducting tape with two solder dots (left) in comparison to a numerical simulation result (right). The simulation shows the temperature of the superconducting tape

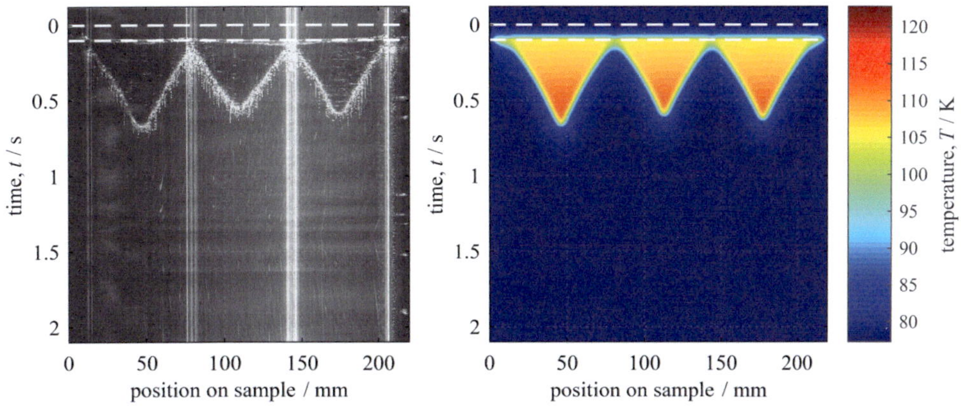

Figure 8.3: Bubble coverage array of a recovering superconducting tape with four solder dots (left) in comparison with a numerical simulation result for the same experimental setup (right)

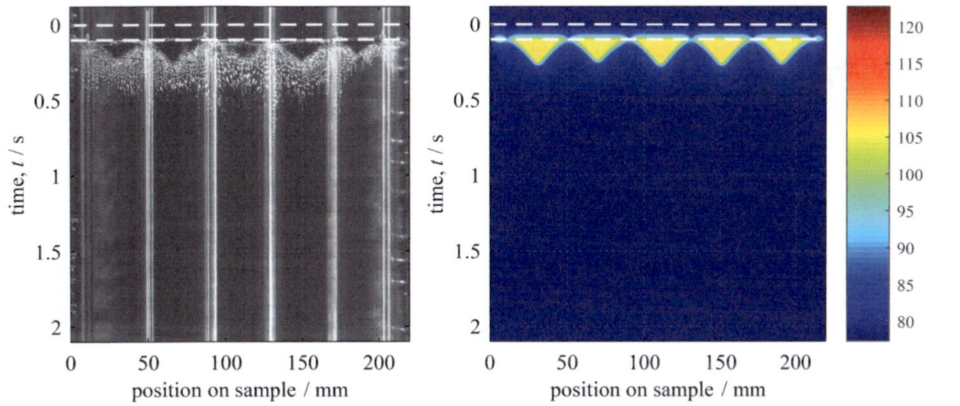

Figure 8.4: Bubble coverage array of a recovering superconducting tape with six solder dots (left) in comparison with a numerical simulation result for the same experimental setup (right)

106.8 K for the tape with six solder dots on the tape surface. The dark blue color at the lower end of the temperature scale in the simulated results represents the recovered tape at a temperature of 77.3 K.

The presented results were published by S. Hellmann et. al. at the Applied Superconductivity Conference 2016.

8.3 Details of the Evaluated Superconducting Tapes

Table 8.1: Specific properties for tape 1 to tape 6 as they were evaluated in Chapter 3.5. Given are the values as stated by the manufacturer whereas the values in parenthesis are measurement results.

property	tape 1	tape 2	tape 3	tape 4	tape 5	tape 6
manufacturer	SuperPower® Inc.	SuNAM® Co. Ltd.	SuperPower® Inc.	SuNAM® Co. Ltd.	SuNAM® Co. Ltd.	SuNAM® Co. Ltd.
critical current	> 140 A (151.4 A)	> 200 A (234.2 A)	109 A (108.5 A)	> 100 A (228 A)	> 100 A (235.2 A)	> 100 A (215.6 A)
critical current variation	± 10 %	± 10 %	± 10 %	± 10 %	± 10 %	± 10 %
critical temperature	- (90.4 K)	- (94.5 K)	- (90.5 K)	- (93.7 K)	- (94.0 K)	- (94.5 K)
copper stabilizer, electroplated	110 µm (105.3 µm)	30 µm (30 µm)	40 µm	40 µm	40 µm	40 µm
stabilizer layer, laminated	-	Copper, 2·100 µm (2·92.9 µm)	-	-	Brass, 2·45 µm	Brass, 2·150 µm
silver layer thickness	2 µm	2 µm	2 µm	2 µm	2 µm	2 µm
substrate thickness	Hastelloy®, 50 µm (48.0 µm)	Stainless steel, 95 µm (95.5 µm)	Hastelloy®, 50 µm	Stainless steel, 95 µm	Stainless steel, 95 µm	Stainless steel, 95 µm
*Re*BCO-layer thickness	1 µm	2 µm	1 µm	2 µm	2 µm	2 µm
total tape thickness	163 µm (160.8 µm)	350 µm (351.5 µm)	93 µm	150 µm	230 µm± 15 µm	450 µm ± 20 µm
tape width	4 mm	4 mm	4 mm	4 mm	4 mm	4 mm
minimum bending radius	< 50 mm	< 50 mm	< 50 mm	-	-	-

8.4 List of Devices for Testing a 1 MVA-Class Current Limiting Transformer

Table 8.2: First part of the complete list of measurement devices used during the tests on the superconducting transformer demonstrator

Device	description	manufacturer	type / model number / description
(D1)	cryogenic temperature sensor	Micro-Measurement, Vishay Precision Group, Inc.	ETG-50D, thin film nickel-foil temperature sensor for cryogenic application
(D2)	precision current source	Knick Elektronische Messgeräte GmbH & Co.	J 45, 0.1 µA - 150 mA
(D3)	universal data acquisition device	National Instruments	USB-6341, 16-bit / 8-channel differential DAQ-device
(D4)	laboratory test transformer	SBA-TrafoTech GmbH	test transformer with multiple secondary voltage taps, Uprim = 400 V, Usec. = 50V - 1000 V, Sn = 400 kVA
(D5)	adjustable power resistor	Ruhstrat GmbH & Co. KG	adjustable single phase / three phase resistor, 0.3 Ω – 1.0 Ω, 1000 V
(D6)	high-current thyristor switch	GvA Leistungselektronik GmbH	1.65 kV / 10 kA, bipolar thyristor switch with optical trigger input and phase-angle synchronization, W01C8999S077
(D7)	transient recorder	Elsys AG	16-channel custom transient recorder, TransAS 2.7.3
(D8)	differential probe	TESTEC Elektronik GmbH	TT-SI 9010, 0-70 MHz, adjustable attenuation: 1:100 / 1:1000
(D9)	Rogowski-coil	Rocoil Ltd.	Rogowski-coil SE 432
(D10)	Rogowski-coil integrator	Rocoil Ltd.	Rogowski-coil integrator IE 776
(D11)	customized Rogowski-coil	KIT	Liquid nitrogen immersible rogowski-coil wound on ABS-winding former
(D12)	high-voltage thyristor switch	GvA Leistungselektronik GmbH	60 kV / 1 kA, bipolar thyristor switch with optical trigger input and phase-angle synchronization, W1C

Table 8.3: Second part of the complete list of measurement devices used during the tests on the superconducting transformer demonstrator

Device	description	manufacturer	type / model number / description
(D13)	customized load resistor	Ruhstrat GmbH	RK-5F-So, 1.73 Ω, 600 kVA, 1000 V(RMS), max. continuous load for 120 s.
(D14)	Rogowski-coil integrator	Rocoil Ltd.	customized Rogowski-coil integrator, IR-3790, IR-3791, IR-3792 and IR-3793
(D15)	universal data acquisition device	National Instruments	USB-6353, 16-bit / 16-channel differential DAQ-device
(D16)	digital volumetric flow meter	ANALYT-MTC GmbH	35816 MLW, 0-100 l / min
(D17)	laser distance sensor	MICRO-EPSILON Messtechnik GmbH & Co. KG	laser optical displacement sensor, ILD1700-200, max. resolution = 12 m at 2.5 kS/s, 670 nm laser
(D18)	short-circuit resistor	manufacturer unknown	25.3 mΩ
(D19)	customized high-voltage divider	KIT	balanced 1:3143 voltage divider for 20 kV to ground measurements

8.5 List of Publications

Paper Publications

1. S. Hellmann, M. Abplanalp, L. Hofstetter, M. Noe. Manufacturing of a 1MVA-Class Superconducting Fault Current Limiting Transformer with Recovery-under-Load Capabilities. *IEEE Transaction on Applied Superconductivity*. Volume 27, Issue 4. 2017

2. S. Hellmann, M. Noe. Influence of Different Surface Treatments on the Heat Flux From Solids to Liquid Nitrogen, *IEEE Transaction on Applied Superconductivity*, Volume 24, Issue 3. 2014

3. J.-H. Fournier Lupien, C. Lacroix, S. Hellmann, F. Sirois. Spatial Modulation of Joule-Losses to Increase the Normal Zone Propagation Velocity in (*Re*)BaCuO-Tapes. (arXiv:1710.05409). 2017

4. V. R. Machavaram, L. Wang, S. D. Pandita, S. Hellmann, F. N. Bogonez, G. F. Fernando. Multi-Point Monitoring of Cross-Linking Reactions, *Journal of Applied*

Polymer Science. Volume 131, Issue 22. 2014

5. R. A. Badcock, N. J. Long, M. Mulholland, S. Hellmann, A. Wright, K. A. Hamilton. Progress in the Manufacture of Long Length YBCO-Roebel Cables, *IEEE Transaction on Applied Superconductivity*. Volume 19, Issue 3. 2009

Oral Conference Proceedings

1. S. Hellmann, M. Abplanalp, L. Hofstetter, M. Noe. Manufacturing of a 1MVA-Class Superconducting Fault Current Limiting Transformer with Recovery-under-Load Capabilities. *Applied Superconductivity Conference 2016*. 4LOr2A-02. Denver. 2016

2. S. Hellmann, M. Noe. Optical and Electrical Investigation of a Novel Method for Improving the Recovery under Load Characteristics of Thin Film Superconductors, *European Conference on Applied Superconductivity 2015*. 3M-LS-O2-7. Lyon. 2015

3. S. Hellmann, M. Noe. Influence of different Surface Treatments on the Heat Flux from Solids to Liquid Nitrogen. *International Conference on Magnet Technology 23*. 5OrBA-04. Boston. 2013

Poster Conference Proceedings

1. S. Hellmann, M. Abplanalp, M. Noe. AC-Loss Measurements and Detailed Loss Analysis on a 1 MVA-Class Superconducting Fault Current Limiting Transformer. *European Conference on Applied Superconductivity 2017*. 3LP6-12. Geneva. 2017

2. S. Hellmann, M. Noe. Optical Method for Improving the Accuracy of Intrinsic Normal Zone Temperature Measurements on Thin Film Superconductors during Recovery under Load. *International Conference on Magnet Technology 24*. 2PoCI_19. Seoul. 2015

3. S. Hellmann, M. Abplanalp, L. Hofstetter, M. Noe. Conceptual Design of a 1 MVA Class Fault Current Limiting Superconducting Transformer. *Applied Superconductivity Conference 2014*. 2LPo2C-06:133. Charlotte. 2014

4. S. Hellmann, A. Kudymow, M. Noe. Influence of different Surface Treatments on the Heat Flux from Solids to Liquid Nitrogen, *European Conference on Applied Superconductivity 2013*. 2P-WT2-05. Genova. 2013

8.6 List of Patents

1. Kurzschlussstrombegrenzende Spulenanordnung mit reluktanzkraftbewegtem Eisenkernsegment. Participation: 100 %. Patent pending. 2017
2. Transformator, Wickelkörper dafür und Verfahren zur Herstellung eines Wickelkörpers (Transformer, Winding Former for Transformer and Manufacturing Process for Transformer Winding Formers). Participation: 100 %. Patent pending. 2016
3. Netzanbindung eines Offshore-Windparks an ein Onshore-Netz (Grid Connection between Offshore Windfarms and Onshore Electric Grids). Participation: 20 %. Patent pending. 2015

8.7 List of Supervised Theses

1. Optische und elektrische Untersuchung des Quench- und Recoveryverhaltens von Dünnschichtsupraleitern. Alain Eicher. Bachelor Thesis. 2015
2. Untersuchung von mechanischen Presskontakten zur Stromeinkopplung in supraleitende *Re*BCO-Dünnschichtbandleiter. Diana Sarah Kalteisen. Master Thesis. 2015
3. Umsetzung einer Messroutine zur effizienten Bestimmung von kritischen Strömen bei Hochtemperatursupraleitern. Leonard Weidemann. Bachelor Thesis. 2015
4. Messung von Quench- und Rückkühlzeiten verschiedener Supraleiteranordnungen in flüssigem Stickstoff. Max Heinrich Görtz. Bachelor Thesis. 2013
5. Experimentelle Ermittlung des Wärmeübergangs an Bandleitern in flüssigem Stickstoff. Moritz Bader. Bachelor Thesis. 2012

Nomenclature

α	angle
β	critical current reduction factor
$\Delta H_{\mathrm{vap,LN_2}}$	enthalpy of vaporization of liquid nitrogen
$\Delta L_{\mathrm{LN_2}}$	change in liquid nitrogen filling level
ΔT	temperature change during simulation or temperature difference
Δt	simulation time step
ΔT_{sc}	temperature change in superconductor
ΔU_{sc}	change of thermal energy in the superconductor
$\dot{q}$	heat-flux density into the liquid nitrogen bath
$\dot{q}_{\mathrm{max}}$	maximum heat transfer
$\dot{Q}_{\mathrm{AC-loss}}$	heat flow into the transformer cryostat caused by AC-losses
$\dot{Q}_{\mathrm{CL}}$	heat flow along the current leads
$\dot{Q}_{\mathrm{contacts}}$	heat flow into the transformer cryostat caused by Joule-heating in the contacts, terminating the superconductor tape stacks
$\dot{Q}_{\mathrm{cooling}}$	heat flux into the nitrogen bath
$\dot{Q}_{\mathrm{cryo}}$	heat flow through the cryostat walls
$\dot{Q}_{\mathrm{Cu}}$	heat flow into the transformer cryostat caused by Joule-heating in copper busbars
$\dot{Q}_{\mathrm{hose}}$	heat flow into the hose, connecting the cryostat to the mass flow meter
$\dot{Q}_{\mathrm{in,0}}$	no-load heat-flow rate
$\dot{Q}_{\mathrm{in,LMA}}$	load dependent heat flow determined in loss-measurement A

$\dot{Q}_{\mathrm{in,LMB}}$	load dependent heat flow determined in loss-measurement B
$\dot{Q}_{\mathrm{total}}$	total heat flow
η_{cooling}	refrigeration system efficiency
η_{total}	total transformer efficiency
Φ_{h}	main magnetic flux
π	mathematical constant pi = 3.141592...
$\rho_{\mathrm{N_2}}$	volumetric mass density of gaseous nitrogen
ρ	volumetric mass density
$\rho_{\mathrm{cond.}}$	resistivity of a conductor material
ρ_{Fe}	volumetric mass density of the iron core material
$\rho_{\mathrm{LN_2}}$	volumetric mass density of liquid nitrogen
$\rho_{\mathrm{materials}}$	resistivity of the materials in the superconducting tape
LN_2	liquid nitrogen
$YBa_2Cu_3O_{7-x}$	yttrium-barium-copper-oxide (superconducting compound)
$\vec{B}_{\mathrm{ext.}}$	vector of the external magneric flux density
$\vec{I}_{\mathrm{transp.}}$	transport current vector
a	nominal turn ratio
$A_{\mathrm{Fe,eff.}}$	effective iron core cross section area
A_{Fe}	iron core cross section area
$A_{\mathrm{LN_2}}$	liquid nitrogen surface area inside the cryostat
A_{p}	cross section area of the conductor in the primary winding
A_{sc}	cross section area of the superconductor
$A_{\mathrm{surf.,sc}}$	surface area of the superconductor adjacent to the liquid coolant
$A_{\mathrm{surf.}}$	surface area

A_s	cross section area of the conductor in the secondary winding
B	magnetic flux density
$B_{ext.,c}$	critical external magneric flux density
$B_{ext.}$	external magneric flux density
B_{max}	maximum magnetic flux density in the iron core
$Bi2223$	bismuth-strontium-calcium-copper-oxide, a superconducting compound in the configuration $Bi_2Sr_2Ca_2Cu_3O_{10}$
$c_{p,materials}$	specific heat of the materials in the superconducting tape
c_p	specific heat
dt	sampling time of data acquisition
E	electric field
E_c	critical electric field
f_{grid}	grid frequency
F_L	Lorentz-force
H	magnetic field
$H_{\sigma,gap}$	magnetic stray field between the transformer windings
$H_{\sigma,p}$	stray field inside the primary transformer winding
$H_{\sigma,s}$	stray field inside the secondary transformer winding
H_{σ}	total magnetic stray field
$h_{c,p}$	height of the conductor in the primary winding
$h_{w,p}$	height of the primary winding
$h_{w,s}$	height of the secondary winding
$I_{lim.,peak}$	peak limitation current
$I_{rec.,peak}$	peak recovery current
$I_{rec.}$	recovery current

$I_{c,rel.}$	critical current of the superconductor relative to the critical current in self field
I_c	critical current of the superconductor
$I_{fault,s}$	fault current in the secondary winding
I_{fault}	fault current or short-circuit current
$I_{lim.}$	limited current
$I_{prosp.}$	prospective current
I_p	nominal primary transformer current
$I_{s,start}$	initial secondary transformer current for the numerical simulation
I_{sample}	measurement current through superconductor sample
I_s	nominal secondary transformer current
$I_{transp.}$	transport current in superconductor
J	current density
$J_{c,aver.}$	average critical current density
J_c	critical current density in the superconductor
$J_{p,nom.}$	nominal current density in the primary winding
$J_{s,nom.}$	nominal current density in the secondary winding
k	thermal conductivity
k	thermal conductivity
k_{Fe}	iron core stacking factor
$L_{\sigma,p}$	primary part of stray inductance
$L_{\sigma,s}$	secondary part of stray inductance
L_{σ}	stray inductance of the transformer
L_h	transformer main inductance
L_{load}	inductance of transformer load

l_p	length of the conductor in the primary winding
m_{Fe}	iron core mass
m_{sc}	superconductor mass
n	n-value; exponent describing the exponential EJ-characteristic of a superonducting material
$n_{p,par.}$	number of parallel conductors in the primary winding
$n_{pick-up}$	number of turns in the pick-up coil
N_p	number of turns in the primary winding
$n_{s,par.}$	number of parallel conductors in the secondary winding
N_s	number of turns in the secondary winding
p	pressure
$p1$ to $p4$	parameters for polynominal fit function
P_c	dielectric losses
$P_{Fe,spec.}$	specific losses of the iron core material
P_{Joule}	heat intake from Joule-heating
$P_{loss,AC}$	AC-losses in the superconducting material
$P_{loss,CL}$	current lead losses
$P_{loss,cryo}$	losses through the cryostat walls
$P_{loss,Fe}$	losses in transformer iron core
$P_{loss,load}$	load dependant transformer losses
$P_{loss,p}$	primary winding losses
$P_{loss,total}$	total transformer losses
R	electrical resistance
R'	resistance load per unit length
R_a	average surface roughness

R_{Fe}	representation of magnetization losses in the iron core in transformer equivalent circuit
r_{gap}	medium radius of the stray gap
R_{load}	resistive transformer load
R_p	resistance of the primary transformer winding
r_p	medium radius of the primary winding
R_{sc}	resistance of the superconductor, resistance of the superconducting transformer winding
R_{short}	short-circuit resistor
r_s	medium radius of the secondary winding
R_{tape}	resistance of the superconductive tape
*Re*BCO	rare-earth-barium-copper-oxide (superconducting compound)
S_n	rated complex nominal power
T	temperature
t	time
T_c	critical temperature of a superconductor
t_{fault}	duration of fault
T_{max}	maximum temperature
$T_{sc,start}$	initial temperature of the superconductor in a numerical simulation
T_{sc}	temperature of the superconductor
T_{tape}	temperature of superconducting tape
th_{Ag}	thickness of silver cap layer in superconducting tape composition
th_{Cu}	thickness of copper stabilizer in superconducting tape composition
th_{gap}	thickness of the stray gap in radial direction
th_{HY}	thickness of Hastelloy® substrate in superconducting tape composition

th_{p}	thickness of the primary winding in radial direction
th_{s}	thickness of the secondary winding in radial direction
U	internal energy
U_{grid}	grid voltage
U_{k}	short-circuit voltage
u_{k}	short-circuit impedance
U_{ph1}	voltage between phase 1 of a three-phase system and ground
U_{ph2}	voltage between phase 2 of a three-phase system and ground
$U_{\mathrm{pick-up}}$	pick-up voltage
U_{p}	primary transformer voltage
U_{r}	rated primary voltage
U_{sc}	thermal energy in the superconductor
U_{s}	secondary transformer voltage
u_{s}	induced secondary voltage
u_{w}	turn-to-turn voltage
V	volume
$V_{\mathrm{N_2}}$	volumetric-flow rate of evaporated LN_2
W_{σ}	energy stored in a magnetic stray field
$w_{\mathrm{c,p}}$	width of the conductor in the primary winding
w_{sc}	width of a superconducting tape
X_{σ}	stray reactance of the transformer
Z_{grid}	complex internal impedance of the power grid
Z_{k}	short circuit impedance of the transformer
$Z_{\mathrm{load,fault}}$	complex transformer load impedance during fault

$Z_{load,nom.}$	nominal complex transformer load impedance
Z_{load}	complex transformer load impedance
$Z_{tot.}$	total impedance of the full equivalent transformer circuit
ABS	acrylonitrile-butadiene-styrene
AC	alternating current
BSCCO	bismuth-strontium-calcium-copper-oxide (superconducting compound)
CAD	computer aided design
CNC	computerized numerical control
Cu	copper (chemical element)
D1 to D19	measurement device 1 to measurement device 19
DC	direct current
DIN	Deutsches Institut für Normung
DyX	delta-wye-vector group connection with $X \cdot 30^{\circ}$ lag between primary and secondary transformer current
GFRP	glass fibre reinforced plastic
HTS	high temperature superconductor
HW	half-wave; half of one period of alternating current
ISO	International Organization for Standardization
LTS	low temperature superconductor
MLI	multi-layer insulation
Nb3Sn	triniobium-tin (superconducting alloy)
NbTi	niobium-titanium (superconducting alloy)
pri.	primary
Pt100	resistive platinum temperature sensor with a resistivity of 100 Ω at 0°C
PV	photovoltaic

PVC	polyvinyl-chloride
RMS	root mean square
RuL	recovery-under-load
SC	superconductor
sec.	secondary
T1 to T8	temperature sensor 1 to temperature sensor 8
Y	yttrium (chemical element)
YBCO	yttrium-barium-copper-oxide (superconductor compound)

Bibliography

[AKY+04] N. Amemiya, S. Kasai, K. Yoda, Z. Jiang, G. A. Levin, P. N. Barnes, and C. E. Oberly. AC-loss Reduction of YBCO Coated Conductors by Multifilamentary Structure. *Superconductor Science and Technology*, 17(12):1464–1471, 2004.

[AMBM98] N. Amemiya, S. Murasawa, N. Banno, and K. Miyamoto. Numerical Modelings of Auperconducting Wires for AC-loss Calculations. *Physica C: Superconductivity*, 310(1-4):16–29, 1998.

[AMBT97] N. Amemiya, K. Miyamoto, N. Banno, and O. Tsukamoto. Numerical Analysis of AC-losses in High-T_c Superconductors based on E-J-Characteristics Represented with n-Value. *IEEE Transactions on Applied Superconductivity*, 7(2):0–3, 1997.

[Bar13] C. Barth. *High Temperature Superconductor Cable Concepts for Fusion Magnets*. KIT Scientific Publishing, Karlsruhe, 2013.

[BBW+05] J. Bock, F. Breuer, H. Walter, S. Elschner, M. Kleimaier, R. Kreutz, and M. Noe. CURL 10: Development and Field-Test of a 10kV / 10MVA Resistive Current Limiter Based on Bulk MCP-BSCCO 2212. *IEEE Transactions on Applied Superconductivity*, 15(2):1955–1960, 2005.

[Ber11] A. Berger. *Entwicklung supraleitender, strombegrenzender Transformatoren*. KIT Scientific Publishing, Karlsruhe, 2011.

[BI93] E. H. Brandt and M. Indenbom. Type-II-Superconductor Strip with Current in a Perpendicular Magnetic Field. *Physical Review B*, 48(17):12893–12906, 1993.

[BK04] W. Buckel and R. Kleiner. *Superconductivity: Fundamentals and Applications*. WILEY-VCH, Weinheim, 2004.

[BM86] J. G. Bednorz and K. A. Müller. Possible High T_c Superconductivity in the Ba-La-Cu-O System. *Zeitschrift für Physik B*, 193(64):267–271, 1986.

[BNK11] A. Berger, M. Noe, and A. Kudymow. Test Results of 60 kVA Current Limiting Transformer With Full Recovery Under Load. *IEEE Transactions on Applied Superconductivity*, 21(3):1384–1387, 2011.

[CFSW03] A. Carlson, J. Furh, G. Schemel, and F. Wegscheider. *Life Testing of Power Transformers*, volume 69. Pro Print, Düsseldorf, 2003.

[CL12] H. M. Chang and S. I. Lee. Optimization of Current Leads for HTS-Elements Immersed in Liquid Nitrogen. *IEEE Transactions on Applied Superconductivity*, 22(3):3–6, 2012.

[DPF+10] R. M. Del Vecchio, B. Poulin, P. T. Feghali, D. M. Shah, and R. Ahuja. *Transformer Design Principles*. CRC Press, Boca Raton, second edi edition, 2010.

[EBW+07] S. Elschner, F. Breuer, H. Walter, M. Stemmle, and J. Bock. HTS-Components for High-Voltage Resistive Current Limiters based on a Magnetic Field Triggered Concept. *IEEE Transactions on Applied Superconductivity*, 17(2):1772–1775, 2007.

[EKB+12] S. Elschner, A. Kudymow, J. Brand, S. Fink, W. Goldacker, F. Grilli, M. Noe, M. Vojenciak, A. Hobl, M. Bludau, C. Jänke, S. Krämer, and J. Bock. ENSYSTROB - Design, Manufacturing and Test of a 3-Phase Resistive Fault Current Limiter based on Coated Conductors for Medium Voltage Application. *Physica C: Superconductivity and its Applications*, 482:98–104, 2012.

[Eki06] J. Ekin. *Experimental Techniques for Low-Temperature Measurements*. Oxford University Press, Oxford, 2006.

[FIT+92] K. Funaki, M. Iwakuma, M. Takeo, N. Takahashi, T. Fujii, T. Nakata, M. Hoshino, E. Shimada, E. Yoneda, and T. Fujioka. Electromagnetic Properties of Large Capacity Superconducting Cables in a 1000 kVA-Class Superconducting Transformer. *Cryogenics*, 32(1):489–492, 1992.

[FIT+97] K. Funaki, M. Iwakuma, M. Takeo, K. Yamafuji, J. Suchiro, M. Hara, M. Konno, Y. Kasagawa, I. Itoh, S. Nose, M. Ueyama, K. Hayashi, and K. Sato. Preliminary Tests of a 500 kVA-Class Oxide Superconducting Transformer Cooled by Subcooled Nitrogen. *IEEE Transactions on Appiled Superconductivity*, 7(2):824–827, 1997.

[Fla93] W. M. Flanagan. *Handbook of Transformer Design & Applications*. McGraw-Hill, Boston, 1993.

[Fra18] Fraunhofer-Institut für Solare Energiesysteme ISE. *Installierte Netto-Leistung zur Stromerzeugung in Deutschland*, 2018 (accessed February 13, 2018). https://www.energy-charts.de/power_inst_de.htm.

[FSS03] P. Fortescue, G. Swinerd, and J. Stark. *Spacecraft Systems Engineering*. Wiley, Sussex, 2003.

[FTLB88] A. Fevrier, J. P. Tavergnier, Y. Laumond, and M. Beckhaled. Preliminary Tests on a Superconducting Power Transformer. *IEEE Transactions on Magnetics*, 24(2):1921–1923, 1988.

[Geo09] P. S. Georgilakis. *Spotlight on Modern Transformer Design*. Springer, London, 2009.

[GFK+09] W. Goldacker, A. Frank, A. Kudymow, R. Heller, A. Kling, S. Terzieva, and C. Schmidt. Status of High Transport Current Roebel Assembled Coated Conductor Cables. *Superconductor Science and Technology*, 22(3), 2009.

[Gri04] F. Grilli. *Numerical Modelling of High Temperature Superconducting Tapes and Cables*. PhD thesis, École polytechnique fédérale de Lausanne, 2004.

[GSB+11] N. D. Glasson, M. P. Staines, R. Buckley, M. Pannu, and S. Kalsi. Development of a 1 MVA 3-phase Superconducting Transformer using YBCO Roebel Cable. *IEEE Transactions on Applied Superconductivity*, 21(3 PART 2):1393–1396, 2011.

[GSJA13] N. D. Glasson, M. P. Staines, Z. Jiang, and N. S. Allpress. Verification Testing for a 1 MVA 3-phase Demonstration Transformer using 2G-HTS Roebel Cable. *IEEE Transactions on Applied Superconductivity*, 23(3), 2013.

[GZP+14] F. Grilli, V. M. R. Zermeno, E. Pardo, M. Vojenciak, J. Brand, A. Kario, and W. Goldacker. Self-Field Effects and AC-Losses in Pancake Coils Assembled From Coated Conductor Roebel Cables. *IEEE Transactions on Applied Superconductivity*, 24(3):1–5, 2014.

[HABK93] S. Hörnfeldt, O. Albertsson, D. Bonmann, and F. König. Power Transformer with Superconducting Windings. *IEEE Transactions on Magnetics*, 29(6):3556–3558, 1993.

[Har06] J. H. Harlow. *Electric Power Transformer Engineering*. CRC Press, Boca Raton, 2006.

[Haz17] D. W Hazelton. 2G HTS Wire Development at SuperPower 2G HTS. Technical report, Barcelona, 2017.

[HEB+12] A. Hobl, S. Elschner, J. Bock, S. Kramer, C. Janke, and J. Schramm. Superconducting Fault Current Limiters - A New Tool for the Grid of the Future. *CIRED 2012 Workshop: Integration of Renewables into the Distribution Grid*, (May):296–296, 2012.

[HKHO11] N. Hayakawa, H. Kojima, M. Hanai, and H. Okubo. Progress in Development of Superconducting Fault Current Limiting Transformer (SFCLT). *IEEE Transactions on Applied Superconductivity*, 21(3):1397–1400, 2011.

[HKK+11] N. Hayakawa, T. Kito, H. Kojima, M. Hanai, and H. Okubo. Analysis of Current Limiting and Recovery Characteristics of Superconducting Fault Current Limiting Transformer (SFCLT) With YBCO Coated Conductors. *IEEE Transactions on Applied Superconductivity*, 21(3):1422–1425, 2011.

[HN14] S. Hellmann and M. Noe. Influence of Different Surface Treatments on the Heat Flux from Solids to Liquid Nitrogen. *IEEE Transactions on Applied Superconductivity*, 24(3), 2014.

[IAM+11] A. Ishiyama, M. Arai, H. Momotari, X. Wang, H. Ueda, T. Saito, Y. Aoki, M. Yagi, T. Machi, and N. Fujiwara. Degradation of YBCO Coated Conductors due to an Over Current Pulse. *IEEE Transactions on Applied Superconductivity*, 21(3 PART 3):3025–3028, 2011.

[IMM+95] T. Ise, Y. Marutani, Y. Murakami, E. Yoneda, and R. Sugawara. Characteristics of a 40 kVA Three Phase Superconducting Transformer and its Parallel Operation with a Conventional Transformer. *IEEE Transactions on Applied Superconductivity*, 5(2):941–944, 1995.

[JAN+06] Z. Jiang, N. Amemiya, M. Nakahata, Y. Iijima, K. Kakimoto, T. Saitoh, and Y. Shiohara. AC-Loss Characteristics of YBCO Coated Conductors With Varying Magnitude of Critical Current. *IEEE Transactions on Applied Superconductivity*, 16(2):85–88, 2006.

[JMMW96] R. C. Johnson, B. McConnell, S. P. Mehta, and M. S. Walker. Status of Superconducting Power Transformer Development. Technical report, Oak Ridge National Laboratory, Tennessee, 1996.

[JS73] R. Jacobsen and R. Stewart. Thermodynamic Properties of Nitroen including Liquid and Vapor Phases from 63 K to 2000 K with Pressures to 10000 Bar. *J. Phys. Chem Ref. Data*, 2(4):757–922, 1973.

[Kal11] S. S. Kalsi. *Applications of High Temperature Superconductors to Electric Power Equipment*. John Wiley & Sons, Inc., Piscataway, 2011.

[Kav01] M. Kaviany. *Principles of Convective Heat Transfer*, volume 109. Springer Science+Business Media, New York, 2001.

[KOH$^+$91] Y. Kito, H. Okubo, N. Hayakawa, Y. Mita, and M. Yamamoto. Development of a 6600 V / 210 V 100 kVA Hybrid-Type Superconducting Transformer. *IEEE Transactions on Power Delivery*, 6(2):816–823, 1991.

[Kom95] P. Komarek. *Hochstromanwendung der Supraleitung*. Teubner Stuttgart, Stuttgart, 1995.

[LCEB11] A. C. Lapthorn, I. Chew, W. G. Enright, and P. S. Bodger. HTS Transformer: Construction Details, Test Results, and Noted Failure Mechanisms. *IEEE Transactions on Power Delivery*, 26(1):394–399, 2011.

[LGD05] J. W. Lue, M. J. Gouge, and R. C. Duckworth. Over-Current Testing of HTS Tapes. *IEEE Transactions on Applied Superconductivity*, 15(2 PART II):1835–1838, 2005.

[MC64] H. Merte and J. A. Clark. Boiling Heat Transfer with Cryogenic Fluids at Standard-, Fractional- and Near-Zero Gravity. *Journal of Heat Transfer*, 86(3):351–358, 1964.

[Mel12] Z. Melhem. *High Temperature Superconductors (HTS) for Energy Applications*. Woodhead Publishing Limited, Cambridge, 2012.

[MGC63] B. Matthias, T. Geballe, and V. Compton. Superconductivity. *Reviews of Modern Physics*, 35(1):1–22, 1963.

[MKMK88] Iwakuma M., Funaki K., Takeo M., and Yamafuji K. Fabrication and Preliminary Test of a 72 kVA Superconducting Four-Winding Power Transformer. *Journal of Cryogenics and Superconductivity Society of Japan*, 23(6):362–368, 1988.

[Moo14] S. H. Moon. HTS Wire Development and Industrialization at SuNAM. Technical report, Barcelona, 2014.

[MW02] N. Magnusson and A. Wolfbrandt. AC-Losses in High-Temperature Superconducting Tapes exposed to Longitudinal Magnetic Fields. *Cryogenics*, 41(2001):721–724, 2002.

[Nor70] W. T. Norris. Calculation of Hysteresis Losses in Hard Superconductors carrying AC: Isolated Conductors and Edges of thin Sheets. *Journal of Physics D: Applied Physics*, 3(4):489–507, 1970.

[NS07] M. Noe and M. Steurer. High-temperature Superconductor Fault Current Limiters: Concepts, Applications, and Development Status. *Superconductor Science and Technology*, 20(3):R15–R29, 2007.

[OKH$^+$09] K. Omura, H. Kojima, N. Hayakawa, F. Endo, M. Noe, and H. Okubo. Current Limiting Characteristics of Parallel-Connected Coated Conductors for High-T_c Superconducting Fault Current Limiting Transformer (HT$_c$-SFCLT). 19(3):1880–1883, 2009.

[OKI$^+$07] H. Okubo, C. Kurupakorn, S. Ito, H. Kojima, N. Hayakawa, F. Endo, and M. Noe. High-T_c Superconducting Fault Current Limiting Transformer (HT$_c$-SFCLT) with 2G Coated Conductors. 17(2):1768–1771, 2007.

[PLR$^+$97] W Paul, M. Lakner, J. Rhyner, P. UnternÃ€hrer, T. Baumann, M. Chen, L. Widenhorn, and A. Guérig. Test of 1.2 MVA High-T_c Superconducting Fault Current Limiter. *Superconductor Science and Technology*, 10(12):914–918, 1997.

[PSJG15] E. Pardo, M. Staines, Z. Jiang, and N. Glasson. AC-Loss Modelling and Measurement of Superconducting Transformers with Coated-Conductor Roebel-Cable in Low-Voltage winding. *Superconductor Science and Technology*, 28(11):114008, 2015.

[RJ89] H. M. Ryan and G. R. Jones. *SF_6 Switchgear.* Peter Peregrinus Ltd., London, 1989.

[Roy10] F. Roy. *Modeling and Characterization of Coated Conductors Applied to the Design of Superconducting Fault Current Limiters.* PhD thesis, École polytechnique fédérale de Lausanne, 2010.

[SBE$^+$16] C. Schacherer, A. Bauer, S. Elschner, W. Goldacker, H. P. Kraemer, A. Kudymow, O. Nackel, S. Strauss, and V. M. R. Zermeno. SmartCoil - Concept of a Full-Scale Demonstrator of a Shielded Core Type Superconducting Fault Current Limiter. *IEEE Transactions on Applied Superconductivity*, 8223(c), 2016.

[Sch09] C. Schacherer. *Theoretische und experimentelle Untersuchungen zur Entwicklung supraleitender resistiver Strombegrenzer*. Universitätsverlag Karlsruhe, Karlsruhe, 2009.

[See98] B. Seeber. *Handbook of Applied Superconductivity*. IOP Publishing Ltd., Bristol, 1998.

[Sei15] P. Seidel. *Applied Superconductivity: Handbook on Devices and Applications*. Wiley-VCH, Weinheim, 2015.

[SGP$^+$12] M. Staines, N. Glasson, M. Pannu, K. P. Thakur, R. Badcock, N. Allpress, P. D'Souza, and E. Talantsev. The Development of a Roebel Cable based 1 MVA HTS Transformer. *Superconductor Science and Technology*, 25(1):014002, 2012.

[SKN08] C. Schacherer, A. Kudymow, and M. Noe. Dissipated Energy as a Design Parameter of Coated Conductors for their Use in Resistive Fault Current Limiters. *IEEE Transactions on Applied Superconductivity*, 97(012193), 2008.

[Sla41] P. G. Slade. *The Vacuum Interruptor - Theory, Design and Application*. CRC Press, Boca Raton, 1941.

[SMWJ02] S. W. Schwenterly, S. P. Mehta, M. S. Walker, and R. H. Jones. Development of HTS power Transformers for the 21st Century: Waukesha Electric Systems, IGC-SuperPower, RG&E, ORNL-SPI- collaboration. *Physica C: Superconductivity and its Applications*, 382(1):1–6, 2002.

[SSLM03] R. Schlosser, H. Schmidt, M. Leghissa, and M. Meinert. Development of High-Temperature Superconducting Transformers for Railway Applications. *IEEE Transactions on Appiled Superconductivity*, 13(2):2325–2330, 2003.

[Thy18] Thyssen-Krupp Steel AG. *Typical Material Properties of Powercore C*, 2018 (accessed February 5, 2018). https://www.thyssenkruppsteel.com/de/produkte/elektroband/elektroband-kornorientiert/powercore-c/typische-werkstoffeigenschaften/content-page-71.html.

[WAT+87] M. K. Wu, J. R. Ashburn, C. J. Torng, P. H. Hor, R. L. Meng, L. Gao, Z. J. Huang, Y. Q. Wang, and C. W. Chu. Superconductivity at 93 K in a new Mixed-Phase Yb-Ba-Cu-O Compound System at Ambient Pressure. *Physical Review Letters*, 58(9):908–910, 1987.

[Whe28] H. A. Wheeler. Simple Inductance Formulas for Radio Coils. *Proceedings of the Institute of Radio Engineers*, 16(10):1398–1400, 1928.

[Wil83] M. N. Wilson. *Superconducting Magnets*. Clarendon Press, 1983.

[WMH15] G. W. Webb, F. Marsiglio, and J. E. Hirsch. Superconductivity in the Elements, Alloys and Simple Compounds. *Physica C: Superconductivity and its Applications*, 514:17–27, 2015.

[XGS+13] Y. Xin, W. Z. Gong, Y. W. Sun, J. B. Cui, H. Hong, X. Y. Niu, H. Z. Wang, L. Z. Wang, Q. Li, J. Y. Zhang, Z. Q. Wei, L. Liu, H. Yang, and X. H. Zhu. Factory and Field Tests of a 220 kV/300 MVA Statured Iron-Core Superconducting Fault Current Limiter. *IEEE Transactions on Applied Superconductivity*, 23(3), 2013.

[XZG+17] A. Xu, Y. Zhang, M. Heydari Gharahcheshmeh, Y. Yao, E. Galstyan, D. Abraimov, F. Kametani, A. Polyanskii, J. Jaroszynski, V. Griffin, G. Majkic, D. C. Larbalestier, and V. Selvamanickam. J_e (4.2 K, 31.2 T) beyond 1 kA/mm^2 of a ~3.2 μm thick, 20 mol% Zr-added MOCVD REBCO Coated Conductor. *Scientific Reports*, 7(1):6853, 2017.

[YTMI91] E. S. Yoneda, I. Tashiro, M. Morohoshi, and D. Ito. Tests on a 30 kVA-Class Superconducting Transformer. *Cryogenics*, 31(7):655–659, 1991.

[Zue98] H. Zueger. 630 kVA High Temperature Superconducting Transformer. *Cryogenics*, 38(11):1169–1172, 1998.

[ZYS+15] Y. Zhu, P. Yi, M. Sun, X. Yang, Y. Zhang, and Y. Zhao. The Study of Critical Current for YBCO Tape in Distorted Bending Mode. *Journal of Superconductivity and Novel Magnetism*, 28(12):3519–3523, 2015.

Karlsruher Schriftenreihe zur Supraleitung

Karlsruher Institut für Technologie (KIT) | ISSN 1869-1765

Herausgeber: Prof. Dr.-Ing. M. Noe, Prof. Dr. rer. nat. M. Siegel

Band 001 **Christian Schacherer**
Theoretische und experimentelle Untersuchungen zur Entwicklung supraleitender resistiver Strombegrenzer. 2009
ISBN 978-3-86644-412-6

Band 002 **Alexander Winkler**
Transient behaviour of ITER poloidal field coils. 2011
ISBN 978-3-86644-595-6

Band 003 **André Berger**
Entwicklung supraleitender, strombegrenzender Transformatoren. 2011
ISBN 978-3-86644-637-3

Band 004 **Christoph Kaiser**
High quality Nb/Al-AlOx/Nb Josephson junctions. Technological development and macroscopic quantum experiments. 2011
ISBN 978-3-86644-651-9

Band 005 **Gerd Hammer**
Untersuchung der Eigenschaften von planaren Mikrowellenresonatoren für Kinetic-Inductance Detektoren bei 4,2 K. 2011
ISBN 978-3-86644-715-8

Band 006 **Olaf Mäder**
Simulationen und Experimente zum Stabilitätsverhalten von HTSL-Bandleitern. 2012
ISBN 978-3-86644-868-1

Die Bände sind unter www.ksp.kit.edu als PDF frei verfügbar oder als Druckausgabe bestellbar.

Band 007 **Christian Barth**
High Temperature Superconductor Cable Concepts for Fusion Magnets. 2013
ISBN 978-3-7315-0065-0

Band 008 **Axel Stockhausen**
Optimization of Hot-Electron Bolometers for THz Radiation. 2013
ISBN 978-3-7315-0066-7

Band 009 **Petra Thoma**
Ultra-fast $YBa_2Cu_3O_{7-x}$ direct detectors for the THz frequency range. 2013
ISBN 978-3-7315-0070-4

Band 010 **Dagmar Henrich**
Influence of Material and Geometry on the Performance of Superconducting Nanowire Single-Photon Detectors. 2013
ISBN 978-3-7315-0092-6

Band 011 **Alexander Scheuring**
Ultrabreitbandige Strahlungseinkopplung in THz-Detektoren. 2013
ISBN 978-3-7315-0102-2

Band 012 **Markus Rösch**
Development of lumped element kinetic inductance detectors for mm-wave astronomy at the IRAM 30 m telescope. 2013
ISBN 978-3-7315-0110-7

Band 013 **Johannes Maximilian Meckbach**
Superconducting Multilayer Technology for Josephson Devices. 2013
ISBN 978-3-7315-0122-0

Band 014 **Enrico Rizzo**
Simulations for the optimization of High Temperature Superconductor current leads for nuclear fusion applications. 2014
ISBN 978-3-7315-0132-9

Band 015 **Philipp Krüger**
Optimisation of hysteretic losses in high-temperature superconducting wires. 2014
ISBN 978-3-7315-0185-5

Band 016 **Matthias Hofherr**
Real-time imaging systems for superconducting nanowire single-photon detector arrays. 2014
ISBN 978-3-7315-0229-6

Band 017 **Oliver Näckel**
Development of an Air Coil Superconducting Fault Current Limiter. 2016
ISBN 978-3-7315-0526-6

Band 018 **Christoph M. Bayer**
Characterization of High Temperature Superconductor Cables for Magnet Toroidal Field Coils of the DEMO Fusion Power Plant. 2017
ISBN 978-3-7315-0605-8

Band 019 **Shengnan Zou**
Magnetization of High Temperature Superconducting Trapped-Field Magnets. 2017
ISBN 978-3-7315-0715-4

Band 020 **Ilya Charaev**
Improving the Spectral Bandwidth of Superconducting Nanowire Single-Photon Detectors (SNSPDs). 2018
ISBN 978-3-7315-0745-1

Band 021 **Juliane Raasch**
Electrical-field sensitive $YBa_2Cu_3O_{7-x}$ detectors for real-time monitoring of picosecond THz pulses. 2019
ISBN 978-3-7315-0786-4

Band 022 **Yingzhen Liu**
Design of a superconducting DC wind generator. 2019
ISBN 978-3-7315-0796-3

Band 023 **Sebastian Hellmann**
Research and Technology Development on Superconducting Current Limiting Transformers. 2019
ISBN 978-3-7315-0804-5